SELECTED WRITINGS
~ OF ~

M. CATHERINE THOMAS

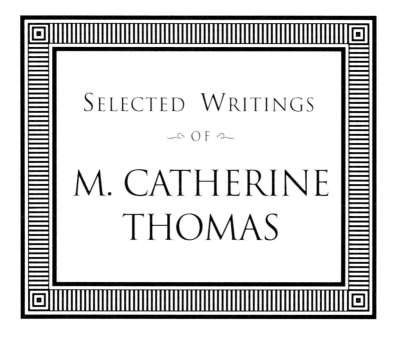

SELECTED WRITINGS

OF

M. CATHERINE
THOMAS

AMALPHI PUBLISHING
PROVO, UTAH

Library of Congress Cataloging-in-Publication Data

Thomas, M. Catherine
  [Selections. 2007]
  Selected writings of M. Catherine Thomas.
    p. cm. — (Gospel Scholars series)
  Includes bibliographical references.
  ISBN 978-1-937735-95-1 (softcover)
  1. Christian life—Mormon authors. I. Title. II. Series.

248.4'893—dc21

00-034650

Printed in the United States of America

72082-6670

10   9   8   7   6   5   4   3   2   1

CONTENTS

PUBLISHER'S PREFACE

In recent decades, a number of exceptional Latter-day Saint scholars have expanded our understanding of many gospel subjects. In many cases, however, much of what they have written (or given as speeches) has been published for relatively small audiences, or it was published so long ago as to be unavailable or inaccessible to most readers; in some cases an article or paper has never been published at all. This collection of some of the "best of the best" was originally published by Deseret Book as part of the Gospel Scholars Series.

KEY TO SHORTENED REFERENCES

Conference Report
   Official reports of general conference sessions of The Church of Jesus
   Christ of Latter-day Saints. Salt Lake City, Utah, 1899–present.

Gospel Doctrine
   Joseph F. Smith. *Gospel Doctrine: Selections from the Sermons and
   Writings of Joseph F. Smith*. Salt Lake City: Deseret Book Co., 1971.

Journal of Discourses
   Journal of Discourses. 26 vols. London: Latter-day Saints' Book
   Depot, 1854–86.

Lectures on Faith
   Joseph Smith, *Lectures on Faith*. Salt Lake City: Deseret Book Co.,
   1985.

Teachings of the Prophet Joseph Smith
   Joseph Smith. *Teachings of the Prophet Joseph Smith*. Selected by
   Joseph Fielding Smith. Salt Lake City: Deseret Book Co., 1976.

Words of Joseph Smith
   Andrew F. Ehat and Lyndon W. Cook, eds., *The Words of Joseph
   Smith: The Contemporary Accounts of the Nauvoo Discourses of the
   Prophet Joseph*. Provo: Religious Studies Center, Brigham Young
   University, 1980.

SCRIPTURE STUDIES

PREMORTAL ELECTION AND GRACE

The world finds God's election of Israel inexplicable, but it would not do for the covenant people themselves to misunderstand their own election. The reason for Israel's election has important implications for us during our mortal probation.

Here is one oft-cited passage from Deuteronomy reflecting God's reasons for the election. Moses proclaimed to Israel: "For thou art an holy people unto the Lord thy God: the Lord thy God hath chosen thee to be a special people unto himself, above all people that are upon the face of the earth. The Lord did not set his love upon you, nor choose you, because ye were more in number than any people; for ye were the fewest of all people: but because the Lord loved you, and because he would keep the oath which he had sworn unto your fathers" (7:6–8). What is the meaning of God's love for Israel?

In his writings the Apostle Paul made several references to Israel's election, indicating that God called Israel in the premortal world and foreordained them to exaltation (Ephesians 1:3–5, 11–12; 2 Timothy 1:9; Romans 8:29; 9:10–12, 17). The well-known passage in Jeremiah speaks on a more individual level: "Before I formed thee in the belly I knew thee; and before thou camest forth out of the womb I sanctified thee" (1:5). Joseph Smith adds to this insight, "Every man who has a calling to minister to the inhabitants of the world was ordained to that very purpose in the Grand Council of heaven before this world was."[1] The Lord has told us very little of what happened in the premortal world, but piecing together some of what has been revealed through scripture and latter-day prophets allows a misty scene to come into sharper focus and provides us with insights that have power to strengthen our faith in the Lord Jesus Christ and to transform our lives.

What can we discover about the events that transpired in the premortal world that have such great implications for us here? Several enlightening scenes appear. In the beginning, the Grand Council of Gods convened to present the plan of salvation to the spirit children. The plan required that man condescend to an earth where his memory of his faculties and glorious condition in the premortal life would be veiled. Elder Parley P. Pratt wrote of the veiling, suggesting also a reduction in man's spiritual condition during mortality: "During man's progress in the flesh, the Holy Spirit may gradually awaken his faculties, and in a dream or vision, or by the spirit of prophecy, reveal, or rather awaken, the memory to a partial vision or to a dim and half-defined recollection of the intelligence of the past. He sees in part and he knows in part, but never while tabernacled in mortal flesh will he fully awake to the intelligence of his former estate. It surpasses his comprehension, is unspeakable, and even unlawful to be uttered."[2]

President Brigham Young remarked on the nature of man's condescension: "It seems to be absolutely necessary in the providence of Him who created us, and who organized and fashioned all things according to his wisdom, that man must descend below all things. It is written of the Savior in the Bible that he descended below all things that he might ascend above all. Is it not so with every man? Certainly it is. It is fit then that we should descend below all things and come up gradually, and learn a little now, and again, receive 'line upon line, precept upon precept, here a little and there a little.'"[3]

Elder John A. Widtsoe wrote, elaborating further on this condescension and perhaps suggesting some apprehension among the spirits about their coming to mortality: "They [mankind] would go to the earth in forgetfulness of the past . . . to be clothed in bodies of 'earth-element,' . . . subject to the conditions of earth, instead of the perfected state of their spirit home. More terrifying was another requirement. Sometime in their earth career their earth-bodies would be separated from their spirit-bodies, a process called death. . . . To subject an eternal being to the dominion of 'earth element'—that is, to forgetfulness, the many vicissitudes of earth, and eventual death—appeared to be a descent in power and station. . . . Man, made to walk upright, must bend his back through the tunnel through the mountain which leads to a beautiful valley. Adam and Eve accepted the call to initiate the plan, and subjected themselves to earth conditions."[4]

Speaking of the reductions that man would experience in this mortal sphere, Elder Joseph F. Smith wrote: "I think that the spirit, before and after this probation, possesses greater facilities, aye, manifold greater, for the acquisition of knowledge, than while manacled and shut up in the prison-house of mortality." President Smith continued: "Had we not known before we came the necessity of our coming, the importance of obtaining tabernacles, the glory to be achieved in posterity, the grand object to be attained by being tried and tested—weighed in the balance, in the exercise of the divine attributes, god-like powers and free agency with which we are endowed; whereby, after descending below all things, Christ-like, we might ascend above all things, and become like our Father, Mother and Elder Brother, Almighty and Eternal!—we never would have come; that is, if we could have stayed away."[5]

President Smith also wrote that while yet in the premortal life, we knew much of what lay before us in mortality, just as the Savior did: "Can we know anything here that we did not know before we came? . . . I believe that our Savior is the ever-living example to all flesh in all these things. He no doubt possessed a foreknowledge of all the vicissitudes through which he would have to pass in the mortal tabernacle, when the foundations of this earth were laid, 'when the morning stars sang together, and all the sons of God shouted for joy.' . . .

"And yet, to accomplish the ultimatum of his previous existence, and consummate the grand and glorious object of his being, and the salvation of his infinite brotherhood, he had to come and take upon him flesh. He is our example. . . . If Christ knew beforehand, so did we. But in coming here, we forgot all, that our agency might be free indeed, to choose good or evil."[6]

Elder Orson Pratt wrote of yet another prospect that would not have appealed to us: "Spirits, though pure and innocent, before they entered the body, would become contaminated by entering a fallen tabernacle; not contaminated by their own sins, but by their connection with a body brought into the world by the fall, earthly, fallen, imperfect, and corrupt in its nature. A spirit, having entered such a tabernacle, though it may commit no personal sins, is unfit to return again into the presence of a holy Being, unless there is an atonement made."[7]

Although we "shouted for joy" (Job 38:7) at the prospect of continuing our progress to our exaltation, is it possible that the contamination of the heavier "earth element," the slower spiritual faculties, the assurance that we would all sin—is it possible that all these perceived sacrifices and reductions produced among the spirit host a sense of vulnerability and fear? Perhaps this is the Apostle Paul's meaning when he seems to say that only our premortal hope in Christ made it possible for us to condescend to the "bondage of corruption," knowing that Christ would deliver us "into the glorious liberty of the children of God" (Romans 8:21).

What was the nature of that hope? As we well know, the divine premortal plan provided for a Savior. "Whom shall I send?" the Lord asked (Abraham 3:27). Jesus volunteered to be that Savior, endorsing the Father's plan. Satan then offered himself—and a radical modification of the plan. The Prophet Joseph Smith described the circumstances: "The contention in heaven was—Jesus said there would be certain souls that would not be saved; and the devil said he could save them all, and laid his plans before the grand council, who gave their vote in favor of Jesus Christ."[8]

As Jesus Christ stood before the spirit congregation in his intelligence and love and majesty, well known to the spirit host, some readily exercised faith and others would not.

We remember that faith is a principle of power in every estate, not just on earth. God himself exercises faith to bring to pass and uphold his creations.[9] Faith in any sphere is a reward for personal righteousness. We may assume that many of the spirit children who exercised faith in Jesus Christ had already had much experience in obeying and searching out divine law in the premortal world, and to them, faith in God's revelations came easily. Thus, two-thirds exercised varying degrees of faith in the yet unrealized promises of Jesus Christ. The Prophet said, "At the first organization in heaven we were all present, and saw the Savior chosen and appointed and the plan of salvation made, and we sanctioned it."[10]

Presumably it was clear to every spirit among those two-thirds that each one's personal success ultimately depended not only on one's own choices but also on the love, power, and constancy of the appointed Savior. It was clear that we would have to rely wholly on the merits of the Savior (2 Nephi 31:19; Moroni 6:4; John 15:5).

We might wonder whether Satan's success with one-third of the spirit host (D&C 29:36) was not a function of fear at the prospect of earth life. Were the necessary reductions and sacrifices so fearful to them that a large portion of the entire spirit host preferred to abort their eternal progression so as not to have to confront earth life in the flesh at all? Did Satan, sensing his political opportunity, capitalize on their fear of the seemingly overwhelming conditions of a fallen world and propose changes in the Father's plan such that no one would have had to exercise faith, rely on grace, or suffer reductions and sacrifice? President John Taylor adds the further insight that Satan wanted to deprive man of his agency so that he, the supposed redeemer, would not have to be subject to man. He wanted to be the glorious Lord without having to suffer what the anointed one would have to suffer.

"Satan . . . wanted to deprive man of his agency, for if man had his agency, it would seem that necessarily the Lord would be subject to him [man]; as is stated, 'For it behooveth the Great Creator that he suffereth himself to become subject unto man in the flesh, and die for all men, that all men might become subject unto him' [2 Nephi 9:5–7].

"The Lord being thus subjected to man, He would be placed in the lowest position to which it was possible for Him to descend; because of the weakness, the corruption and the fallibility of human nature. But if man had his free agency, this necessarily would be the result, and hence, as it is said, Jesus descended below all things that He might be raised above all things; and hence also, while Satan's calculation was to deprive man of his free agency, and to prevent himself or the Only Begotten from being subject to this humiliation and infamy, the Lord's plan was to give man his free agency, provide a redeemer, and suffer that redeemer to endure all the results incidental to such a position, and thus, by offering himself as a substitute and conquering death, hell and the grave, he would ultimately subjugate all things unto himself; and at the same time make it possible for man to obtain an exaltation that he never could have had without his agency."[11]

Whatever the issues, a terrible war ensued. President Brigham Young taught with respect to that war that "the Lord Almighty suffered this schism in heaven to see what his subjects would do preparatory to their coming to this earth."[12] It seems that the Lord allowed the conflict as a sifting of and a preparation for his spirit children. We gained experience in that conflict that would work to our advantage in our mortal probation and shape our earthly experience.

Out of that two-thirds of the spirit children who supported the Lord Jesus Christ's messiahship, it seems that a smaller group distinguished themselves by their valiance in the war in heaven over the issues concerning the plan of salvation (Alma 13:3) and by their exceeding faith in the premortal promises of the Lord Jesus Christ, and that this valiant group became the covenant people, or premortal Israel. Elder Bruce R. McConkie explained: "Israel is an eternal people. She came into being as a chosen and separate congregation before the foundations of the earth were laid; she was a distinct and a peculiar people in preexistence, even as she is in this sphere. Her numbers were known before their mortal birth, and the very land surface of the earth was 'divided to the nations [for] their inheritance' (Deuteronomy 32:8.)"[13]

Elder Melvin J. Ballard said that premortal Israel was "a group of souls tested, tried, and proven before they were born into the world. . . . Through this lineage were to come the true and tried souls that had demonstrated their righteousness in the spirit world before they came here."[14] Alma explained: "This is the manner after which they were ordained—being called and prepared from the foundation of the world according to the foreknowledge of God, on account of their exceeding faith and good works; in the first place [the premortal world] being left to choose good or evil; therefore they having chosen good [e.g., the Lord Jesus Christ and the plan of salvation], and exercising exceedingly great faith, are called with a holy calling" (Alma 13:3).

Because of their exceeding faith in the Lord Jesus Christ and their desires to promote his work on the earth, this premortal group of covenant people were elected to receive extensive grace on the earth in order to accomplish their own Christlike missions. I suggest that this is one meaning behind the Apostle Paul's use of the term "election of grace" (Romans 11:5); that is, that premortal Israel was elected to receive extensive grace in their mortal probation in order to develop and serve their brothers and sisters in Christlike ways.

It might be helpful to point out a distinction here in the possible meanings of *grace*. The LDS Bible Dictionary says that grace means divine help, necessary help, divine enabling power that one cannot provide for oneself.[15] Sometimes, as with resurrection, grace is given regardless of worthiness; but much grace is given based on faith and faithfulness.

The issue of grace brings us to the main question of this chapter: What indeed were the promises of the Lord to his spirit brothers and sisters that gave us courage to come to earth? They are all comprehended in the word *grace*. We are born without a conscious memory of these promises, but the Lord has supplied scripture and personal revelation to restore in some measure our premortal memories so that we can act in faith in this mortal probation as we did in the premortal world.

Amidst the premortal events, as we have seen, the covenanting spirits were foreordained to exaltation. Calling and election, based on faith in the Lord Jesus Christ, was extended to the covenant people. Paul speaks of this premortal promise in this passage: "God hath from the beginning chosen you to salvation [or exaltation] through sanctification of the Spirit and belief of the truth: Whereunto he called you by our gospel, to the obtaining of the glory of our Lord Jesus Christ" (2 Thessalonians 2:13–14). As Paul teaches here, the calling and election to exaltation made in the premortal world was not to be automatically granted, but would come "through sanctification of the Spirit." That is, the tabernacled spirits would have to make their calling and election sure on earth by standing in their premortal covenant with Jesus Christ through baptism, receiving the Holy Ghost, and pursuing sanctification until they received the more sure word of prophecy (D&C 131:5).

It is important to note that *foreordination* does not imply the Calvinistic notion of *predestination,* which term acknowledges no necessary worthiness on the part of those predestined to be saved. Fulfillment of premortal foreordinations is based on obedience to eternal law. One foreordained could choose not to obey and could at any point refuse to accept the premortal plan, fall short of his promised glory, and fail to make his premortal calling and election sure.

Consider this next passage not only for its promise but also for its end result: "And we know that all things work together for good to them that love God, to them who are the called according to his purpose. For whom he did foreknow, he also did predestinate [in Greek, *foreordain*] to be conformed to the image of his Son" (Romans 8:28–29). In this passage Paul teaches the working together for good, or an orchestration of events, for those who love God and were premortally called to assist in the great work of the salvation of men. This promise of orchestration of life events must have been among the most compelling for us. It might be restated in this way: a plan with strongly programmed elements would operate for each covenant person, so that each could have the opportunities necessary to save himself or herself through the continually accessible grace of Christ and, in addition, have a saving influence on as many other people as possible (D&C 86:8–11; 103:9–10). These programmed elements include two main forms of grace: a chosen lineage and the presence of the Lord during our mortal experiences.

Chosen lineage, time, place, family, and so forth. That some kind of organizing power brooded over the families of the earth is clear when one reads that Adam prophesied concerning all the families of the earth, and Enoch beheld them (Moses 5:10; 7:45).

President Harold B. Lee taught: "You have been blessed to have a physical body because of your obedience to certain commandments in that premortal state. You are now born into a family to which you have come, into the nations through which you have come, as a reward for the kind of lives you lived before you came here and at a time in the world's history, as the Apostle Paul taught the men of Athens [Acts 17:26] and as the Lord revealed to Moses, determined by the faithfulness of each of those who lived before this world was created."[16]

The covenant spirits would be born through a chosen lineage that has the right to the priesthood, an essential and powerful form of grace. This lineage, of course, began with Adam but was carefully funneled through subsequent selected lines until it came to Abraham, Isaac, and then Jacob, who was renamed Israel. Covenant Israel, or the members of the Church of Jesus Christ today, possess the literal blood of Abraham, Isaac, and Jacob in their veins (Abraham 2:10–11; D&C 86:8–11). Some have called this "believing blood."

Not only was the lineage selected, but also the time and place, the family, and the order of birth, so that each spirit would have that unique configuration of experiences necessary for his or her particular divine needs, including membership in the Church. This careful placement of a person into the temporal house of Israel is part of the election of grace (D&C 84:99; Romans 11:5–7; LDS Bible Dictionary, "Grace").

In connection with this grace of chosen lineage and birth circumstances, God provided yet an additional combination of blessings to give every spirit the greatest possible power to obtain exaltation. I want to focus closely now on one specific, also very encouraging and liberating, form of grace. The spirit of it is captured in these words: Jehovah says to Abraham, "I know the end from the beginning; therefore my hand shall be over thee" (Abraham 2:8).

Individually orchestrated life events. I shall address this section somewhat more personally. I catch myself saying in this telestial world, "I am afraid of this happening," or "What if that happens, then such and such will happen." I hear my loved ones projecting fear into the future and I realize that some of us are either just uninformed concerning our privileges as covenant people or are spiritually careless with what we do know. So I am advocating an *informed* faith—not just wishful thinking that hopes that everything will work out okay, but real faith based on an understanding of the express teachings of the Lord Jesus Christ.

There is a troublesome question that the premortal covenant of grace answers: Is the universe random? Can just anything happen to us or our loved ones when random events join together to produce catastrophe and we are caught in the middle? As we sense our personal vulnerability in a world of overwhelming forces, we are susceptible to a telestial view that the whole cosmos is out of control and that just anything *could* happen. When we are operating under a telestial perception, we can easily feel victimized by the seemingly uncontrollable powers of nature and the malignant agency of men and by that terrifying word *accident*. It can seem as though we were cast out of the premortal existence like so many random dice into an earthly game of "whose agency could ruin whose life?" We may even fear that the great God of heaven and earth wound up the whole creation like a clock and either went off to do something else while we ran down, or at best is only intermittently accessible. Or we fear that perhaps he has sat around, restrained by some omnipotent law of agency, helplessly beholding the mess that man would make.

But let us, who tend to get caught up in patterns of worry,

reconsider three principles from scripture: God is perfectly omniscient, he is omnipotent, and he loves his children with a perfect love. These three facts assure us that he has both the love and the power to restore us to himself, having led and empowered us through an orchestrated course of events in our mortal probation over which he has perfect control.

God is perfectly omniscient. Many scriptures make it abundantly clear that there is not anything that God does not know (2 Nephi 9:20; 27:10; D&C 130:7; Moses 1:6, 27–28). Elder Neal A. Maxwell observed: "God's omniscience is *not* solely a function of prolonged and discerning familiarity with us—but of the stunning reality that the past and present and future are part of an 'eternal now' with God! (Joseph Smith, *History of the Church,* 4:597.) . . . [F]or God to foresee is not to cause or even to desire a particular occurrence—but it is to take that occurrence into account beforehand, so that divine reckoning folds it into the unfolding purposes of God. . . . God has foreseen what we will do and has taken our decision into account (in composite with all others), so that His purposes are not frustrated."[17]

God is omnipotent. He has power to prevent or produce any event. The Lord's power is "over all the inhabitants of the earth" (1 Nephi 1:14); "the Lord is able to do all things according to his will" (1 Nephi 7:12); the Lord has "all power unto the fulfilling of all his words" (1 Nephi 9:6); "by the power of his almighty word [God] can cause the earth [to] pass away" (1 Nephi 17:46); God is "the all-powerful Creator of heaven and earth" (Jacob 2:5); and so on. *Lectures on Faith* reinforce God's power over all things to make salvation possible: "Unless God had power over all things and was able, by his power, to control all things, and thereby deliver his creatures who put their trust in him from the power of all beings that might seek their destruction, whether in heaven, on earth, or in hell, men could not be saved."[18]

God loves his children with a perfect love. In other words, he is emotionally committed to his children. "For behold, this is my work and my glory—to bring to pass the immortality and eternal life of man" (Moses 1:39). God has invested his godly heart and his mighty

resources in causing things to work together for man's greatest blessing; in fact, "He doeth not anything save it be for the benefit of the
world; for he loveth the world, even that he layeth down his own life
that he may draw all men unto him" (2 Nephi 26:24).

From the foregoing truths, that God is omniscient, omnipotent,
and fully committed to us, we may conclude as follows: The fact that
mankind's agency operates on the earth does not require a creation
that is out of control. The truth is that men choose what they will do,
but God limits or extends or orchestrates their choices into his great
overarching purposes for his children. He works through natural and
spiritual law to see that his purposes are fulfilled. Man's choices are
necessarily restricted in order to protect the universe; still, man has
that amount of agency necessary either to be damned or to be
exalted.

Man's choices are circumscribed by the type of kingdom he lives
in. He must remain within the bounds in which the Lord has placed
him (D&C 88:38). He exceeds telestial bounds only through exercising terrestrial or celestial principles. We are invited to exceed our
telestial bounds through making a connection with the laws and
powers of Jesus Christ.

When a person makes a covenant with the Lord Jesus Christ to
obey his commandments and to strive for true and honest discipleship, a whole set of transcendent laws comes into play for that
person: "That which is governed by law [the law of Christ] is also
preserved by law and perfected and sanctified by the same. That
which breaketh a law [the law of Christ], and abideth not by law, but
seeketh to become a law unto itself, and willeth to abide in sin, and
altogether abideth in sin, cannot be sanctified by law, neither by
mercy, justice, nor judgment" (D&C 88:34–35). Heavenly Father's
laws are very powerful and are a great gift to us, because when we
take care to obey them, these divine laws are activated to preserve us
and prosper us (Psalm 37:23; 121:4, 8). Our obedience to the laws
and ordinances of the gospel lifts us above a seemingly random-event
universe and puts us in a situation where we may dwell safely in the
Holy One of Israel (1 Nephi 22:28). But he warns: "I, the Lord, am

bound when ye do what I say; but when ye do not what I say, ye have no promise" (D&C 82:10).

However, as we well know, election to grace will not preclude bad things from happening to good people. Nephi says of himself in the very first verse of the Book of Mormon, "having seen many afflictions in the course of my days, nevertheless, [I have] been highly favored of the Lord in all my days" (1 Nephi 1:1). Even though, as part of our divine curriculum, we are called to go through different kinds of affliction (D&C 122:5–7), these are often lightened by the Lord because of our covenant relationship with him. The Lord speaks to Alma's people in their afflictions: "Lift up your heads and be of good comfort, for I know of the covenant which ye have made unto me. . . . I will . . . ease the burdens which are put upon your shoulders, that even you cannot feel them upon your backs, even while ye are in bondage; and this will I do that ye may stand as witnesses for me hereafter, and that ye may know of a surety that I, the Lord God, do visit my people in their afflictions" (Mosiah 24:13–14; see also Alma 33:23; 38:5).

With this informed and restored faith, we, the covenant people, can rest, like Jesus, while the tempest rages (Matthew 8:24). God will "wake us" if we need to be wakened; God will bring to our attention that which it is necessary for us to know when we need to know it. He may also withhold certain information so as to precipitate a situation which is pregnant with learning possibilities for us. The right orchestration, the right timing for events, even sad ones, is part of the covenant promise—if we hold up our end of the covenant as best we can. Maybe we can even give up the phrase, "If I had only known!"

Even one's appointed days on the earth are known beforehand by the Lord. Elder Neal A. Maxwell wrote: "The lifespans of planets, as well as prophets, are known to God; the former pass away by his word. (Moses 1:35.) To a suffering Joseph Smith, God said, 'Thy days are known, and thy years shall not be numbered less.' (D&C 122:9.) Such a promise could not have been made if all other things that bore upon the lifespan of Joseph Smith were not also known beforehand to God—in perfectness."[19] The Psalmist wrote: "All the days ordained for me were written in your book before one of them came to be" (New International Version, Psalm 139:16; see also Alma 40:10; Helaman 8:8; D&C 42:48; 63:3; 121:25; Acts 17:26; Isaiah 38:5). President Ezra Taft Benson, speaking at President Spencer W. Kimball's funeral, said: "It has been said that the death of a righteous man is never untimely because our Father sets the time. I believe that with all my soul."[20]

When we covenant with the Lord Jesus Christ, we become passengers on a plan that moves steadily toward its destiny in the celestial kingdom, with all the sights, sensations, relationships, and experiences particularly selected for our personal, most effective use of this preparatory period. Elder Richard G. Scott said: "He would have you suffer no consequence, no challenge, endure no burden that is superfluous to your good."[21] We may know then for ourselves that what is coming in our individual futures is benign in its design if we will pay attention to the lesson it means to impart while we are faithful to the Lord.

Elder John A. Widtsoe wrote about the place of chance in a perfectly ordered universe: "In my life's adventure, there has been no chance. Indeed, I doubt if chance has any place in a world supervised by a divine intelligence. Therefore, I have felt that the power from the unseen world has ever been over me and directing my life's course. That faith has removed both fear and dissatisfaction, enemies of mankind. Certainty comes to dwell when chance is removed. It has been easy to approach God in all my work."[22]

Therefore we can go about our business and not fear that some random event will arrest our forward progress to our spiritual destiny (D&C 98:14; 122:9). The Lord reassures us: "The works, and the designs, and the purposes of God cannot be frustrated, neither can they come to naught. . . . Remember, remember that it is not the work of God that is frustrated, but the work of men" (D&C 3:1, 3).

One might be tempted to think that if God knows everything that we will do, if the plan is already in place, maybe we can just quit trying and let everything unfold without our own effort. But we must keep in mind that, in some way that is perhaps beyond the finite mind to understand, two things are true at once: God knows it all, but we must strive to our very utmost to make the plan work for our eternal development. The principle is something like the axiom that we must pray as though everything depended on the Lord, but work as though everything depended on us.

Latter-day Saints are premortal Israel, come to earth. It was our premortal faith in the comprehensive promises of Jesus Christ to Israel that gave us the courage to come to this life in the first place. Our faith in our election to grace can replace our telestial fears. The Apostle Paul reminds Israel that "the just shall live by faith" (Romans 1:17). And with his great faith, the Prophet Joseph Smith gave the Saints this counsel: "Therefore, dearly beloved brethren, let us cheerfully do all things that lie in our power; and then may we stand still, with the utmost assurance, to see the salvation of God, and for his arm to be revealed" (D&C 123:17).

Notes

From *Spiritual Lightening* (Salt Lake City: Bookcraft, 1996), 31–46.

1. *Teachings of the Prophet Joseph Smith,* 365.
2. Parley P. Pratt, *Key to the Science of Theology* (Salt Lake City: Deseret Book Co., 1978), 31.
3. Brigham Young, in *Journal of Discourses,* 15:3.
4. John A. Widtsoe, *Evidences and Reconciliations,* comp. G. Homer Durham (Salt Lake City: Bookcraft, 1960), 73–74.
5. *Gospel Doctrine,* 13.
6. Ibid.
7. Orson Pratt, "The Pre-existence of Man," *The Seer,* July 1853, 98; see also Mosiah 3:16.
8. *Teachings of the Prophet Joseph Smith,* 357.
9. Hebrews 11:3; this scripture and the principle of God's exercising faith to bring to pass the Creation are explained in *Lectures on Faith,* 1:14–17.
10. *Teachings of the Prophet Joseph Smith,* 181.
11. John Taylor, *The Mediation and Atonement* (Salt Lake City: Deseret News Co., 1882), 142.
12. Brigham Young, in *Journal of Discourses,* 14:93.
13. Bruce R. McConkie, *A New Witness for the Articles of Faith* (Salt Lake City: Deseret Book Co., 1985), 510–11.
14. Melvin R. Ballard, *Melvin J. Ballard: Crusader for Righteousness* (Salt Lake City: Bookcraft, 1966), 218–19.
15. See LDS Bible Dictionary, "Grace," 697.

16. Harold B. Lee, "Understanding Who We Are Brings Self-Respect," *Ensign,* January 1974, 5.
17. Neal A. Maxwell, *All These Things Shall Give Thee Experience* (Salt Lake City: Deseret Book Co., 1979), 8, 12.
18. *Lectures on Faith,* 4:12.
19. Maxwell, *All These Things Shall Give Thee Experience,* 13.
20. Ezra Taft Benson, "Spencer W. Kimball: A Star of the First Magnitude," *Ensign,* December 1985, 33.
21. Richard G. Scott, "Obtaining Help from the Lord," *Ensign,* November 1991, 86.
22. John A. Widtsoe, as quoted by Ray H. Wood, "Encouraging advice prophetic for couple embarking on future," *Church News,* 11 July 1998, 5.

Chapter 2

THE YOUNGER ALMA AND THE FATHERS

In Alma 13 the prophet brushes only lightly but so provocatively the major issues of man's existence. In this chapter I would like to explore three main themes touched on by Alma: the premortal arrangements of the fathers, how these affect our families and purposes now, and how these arrangements affect our opportunities after this life.

PREMORTAL ARRANGEMENTS AND THE HOLY CALLING

Alma uses the phrases "from the foundation of the world" and "in the first place" to refer to premortal events. He speaks in particular of a group of people receiving a holy calling in the premortal world (Alma 13:3). To understand more about the holy calling, let us briefly review the premortal organization of the house of Israel.

Out of all of Heavenly Father's spirit children, a smaller group distinguished itself by its exceeding faith in the Lord Jesus Christ during the conflicts incident to the war in heaven. Those who were valiant in these conflicts demonstrated both their abilities and their desires to participate in the cosmic work of redemption through the great atonement of the Lord Jesus Christ. They were issued an invitation, which might also be termed a calling, to become part of the lineage of the house of Israel in heaven and later on earth. Today, on earth, those who now are members of the Church constitute the covenant people of the Lord or the house of Israel. Of course, many more will be reclaimed from the nations of the earth and joined with the covenant people, for it is Israel we are seeking through missionary work. These people were called in the premortal world and, having been channeled through a chosen lineage, are literally blood Israel. Making reference to an Old Testament passage, Joseph Fielding Smith wrote:

"The numbers of the children of Israel were known and the bounds of their habitation fixed, in the days of old when the Lord divided to the nations their inheritance. We conclude, therefore, that there must have been a division of the spirits of men in the spiritual world, and *those who were appointed to be the children of Israel were separated and prepared for a special inheritance.*"[1]

With respect to the literal blood lineage of Israel on earth, Elder Bruce R. McConkie emphatically taught:

"We are literally of the seed of Abraham. Let's just drill it into ourselves! We are literally of the seed of Abraham. We are natural heirs according to the flesh. We are not adopted or anything else. I don't know how there could be language more express than these revelations, 'natural heirs according to the flesh,' 'lawful heirs,' 'the literal seed of the body' [Abraham 2:11]. You see [the Lord] just goes out of his way to make it literal. The literal seed of your body has the right to the priesthood and the gospel, and that is us. Now, granted that somebody can be adopted in, but they are so few and far between up to now, that we can just about forget about them."[2]

Membership in Israel offers certain privileges. When Abraham's grandson, Jacob, was visited by the Lord, the Lord said to him, "Thy name shall be called no more Jacob, but Israel: for as a prince hast thou power with God and with men, and hast prevailed" (Genesis 32:28). *Israel,* then, connotes that the chosen lineage have power and influence with God to bring to pass holy purposes on the earth. Every member of Israel has a holy purpose.

These promises fully include the women of Israel. Abraham's wife was renamed *Sarah* by the Lord, which means "princess," suggesting that all these holy men had holy female counterparts; out of deference to their holiness, most of these women are not even named.

Learning that we are members of the house of Israel because we were invited into that lineage tells us a good deal about ourselves in the former world. This knowledge will help us retrieve our true identities as well as discern the work that has been set out for each of us here on earth. The scriptures and revelations "set forth that there are men [and women] pre-appointed to perform certain works in their lifetime, and bring to pass certain ends and purposes in the economy of heaven."[3] Each of us has a role to play in the interweaving purposes of the heavenly economy.

One thing that a person can learn about himself is that he loved the work of laboring with his fellow beings to bring to pass their advancement. It is this very love that characterizes the gods, whose ceaseless endeavors pertain to the work of advancement and exaltation (Moses 1:39).

If we could look into the heavens, we would likely be stunned at the magnitude of the great work of redemption going on in numberless worlds throughout the cosmos. The work of redemption, in fact, is the major reason that worlds are created and inhabited (1 Nephi 17:36). In fact, redemption is not just one of the things going on in the universe, it is *the* thing. That work of redemption is *the* work to which the premortal covenant people, the house of Israel, were called, and it was to take precedence over all other work and to subordinate all other work to itself. More specifically, the great work of the gods has to do with family work, the nurturing and advancing of children and the redemption and sealing of families, generation after generation. We cannot comprehend with our finite minds the cosmic proportions of the love and the infinite investment of labor and grace that go into this magnificent work—nor how great the privilege to engage in this work ourselves. Again, it was the very desire for that work that elicited our invitation into the royal house of Israel.

All who entered into the premortal house of Israel to participate in the work of redemption were foreordained to be conformed to the image of the son of God (Romans 8:29). That is, those in the premortal world who elected to become gods elected also to come to earth and learn the work of redemption in apprenticeship to the Lord Jesus Christ, a work that would qualify them to live with the gods in the eternal worlds.

As people do in this world, the maturing spirits in the premortal world advanced at different rates and to different levels of spirituality, though Alma says that "in the first place" they all were on the same standing with each other (Alma 13:5); that is, they had equal opportunity to advance; but some rejected the Spirit of God on account of the hardness of their hearts and blindness of their minds (Alma 13:4) and did not make the progress that would have given them the privilege of a greater calling. Many members of Israel embraced opportunities to grow in light and truth and emerged as leaders within the house of Israel. These were prepared and ordained to the holy priesthood order, which was the order of the Son of God, to show the manner of "Christness" to the people, they being types of Christ themselves (Alma 13:16; D&C 138:53–56). Through the ministry of these high priests, the people would understand what they themselves had been foreordained to become.

These high priests began their redeeming labors among the spirits in the premortal world and were ordained and prepared to descend to earth and be leaders in the Lord's work here. We can assume that many of us in the house of Israel had positions of leadership and participated in redeeming work even before this life. The Prophet Joseph, as one of those premortal leaders, taught about the premortal calling of priesthood holders:

"Every man who has a calling to minister to the inhabitants of the world was ordained to that very purpose in the Grand Council of heaven before this world was. I suppose I was ordained to this very office in that Grand Council."[4]

The following verses in Alma reflect what we have just discussed about the holy calling: "I would that ye should remember that the Lord God ordained priests, after his holy order, which was after the order of his Son, to teach these things unto the people. And those priests were ordained after the order of his Son, in a manner that thereby the people might know in what manner to look forward to his Son for redemption.

"And this is the manner after which they were ordained—being called and prepared from the foundation of the world according to the foreknowledge of God, on account of their exceeding faith and good works; in the first place [premortal world] being left to choose good or evil [conflict in heaven]; therefore they having chosen good, and exercising exceedingly great faith, are called with a holy calling, yea, with that holy calling which was prepared with, and according to, a preparatory redemption for such. . . .

"And thus being called by this holy calling, and ordained unto the high priesthood of the holy order of God, to teach his commandments unto the children of men, that they also might enter into his rest" (Alma 13:1–3, 6).

BLESSINGS OF THE FATHERS

"The noble and great ones," like Abraham and many, many others, were among those who distinguished themselves in the premortal world. These were they whom the gods would prove to see if they would do all things whatsoever the Lord their God should command them (Abraham 3:22–23, 25). They had kept their first premortal estate; if they kept their second estate, their earthly probation, they would have glory added upon their heads for ever and ever (Abraham 3:25–26). These are the ones for whom the earth was created (Abraham 3:24) and for whom the scriptures were written (D&C 35:20). These are they of whom Jesus said, "All that the Father giveth me shall come to me. . . . And this is the Father's will which hath sent me, that of all which he hath given me I should lose nothing, but should raise it up again at the last day" (John 6:37, 39). These are they who could be characterized in the premortal existence by these words of Abraham as he expressed his primeval desire to bring himself and others to godliness as the fathers before him had done:

"I sought for the blessings of the fathers, and the right whereunto I should be ordained *to administer the same;* having been myself a follower of righteousness, desiring also to be one who possessed great knowledge, and to be a greater follower of righteousness, and to possess a greater knowledge, and to be a *father* of many nations, a *prince* of peace, and desiring to receive instructions, and to keep the commandments of God, I became a rightful heir, a High Priest, holding the right belonging to the fathers. It was conferred upon me from the fathers; it came down from the fathers, . . . from the beginning, or before the foundation of the earth, down to the present time, even the right of the firstborn, or the first man, who is Adam, or first father, through the fathers unto me" (Abraham 1:2–3).[5]

That desire which characterizes Israel, then, is the desire to have a saving influence through the power of the holy priesthood on all our connections—to be the mothers and fathers and nurturers of saved beings.

There are very specific reasons for the revelation of premortal events: "Most of us have wondered about what occurred in the premortal world and how it relates to our existence here. We should be acquainted with the truth that knowledge of the premortal life was restored that we might fulfill our responsibilities as children of God. . . .

"John A. Widtsoe provides insight to an earth-life responsibility made in that premortal world which is of great importance. He highlights a contractual agreement we made concerning the eternal welfare of all of the sons and daughters of the Eternal Father: ' . . . Since the plan is intended for all men, we [the covenant people] became parties to the salvation of every person under that plan. We agreed, right then and there, to be not only saviors for ourselves but . . . saviors for the whole human family. We went into a partnership with the Lord. . . . The least of us, the humblest, is in partnership with the Almighty in achieving the purpose of the eternal plan of salvation.'"[6]

In order to prepare our minds to participate in this work, many of the Brethren have spoken on the premortal preparations and assignments of the house of Israel. Here is a sampling:

Joseph Fielding Smith: "During the ages in which we dwelt in the pre-mortal state we not only developed our various characteristics and showed our worthiness and ability, . . . we were also where such progress could be observed. It is reasonable to believe that there was a Church organization there. The heavenly beings were living in a perfectly arranged society. Every person knew his place. Priesthood . . . had been conferred and the leaders were chosen to officiate. Ordinances pertaining to that pre-existence were required and the love of God prevailed."[7]

Wilford Woodruff: "The Lord has chosen a small number of choice spirits of sons and daughters out of all the creations of God, who are to inherit this earth; and this company of choice spirits have been kept in the spirit world for six thousand years to come forth in the last days, to stand in the flesh in this last Dispensation of the Fulness of Times, to organize the Kingdom of God upon the earth, to build it up and to defend it . . . and to receive the eternal and everlasting Priesthood."[8]

Spencer W. Kimball: "Remember, in the world before we came here, faithful women were given certain assignments while faithful men were foreordained to certain priesthood tasks. While we do not now remember the particulars, this does not alter the glorious reality of what we once agreed to. You are accountable for those things which long ago were expected of you just as are those we sustain as prophets and apostles!"[9]

The Prophet Joseph taught of an organization of families in the premortal world and points to the eternal potential of that organization: "At the first *organization* in heaven we were all present, and saw the Savior chosen and appointed and the plan of salvation made, and we sanctioned it."[10] What was the first organization? As Joseph Smith said to Brigham Young in a dream in 1847: "Be sure to tell the people to keep the Spirit of the Lord; and if they will, they will find themselves just as they were organized by our Father in heaven, before they came into the world. Our Father in heaven organized the human family, but they are all disorganized and in great confusion."[11]

President Young continued his record: "Joseph then showed me the pattern, how they were in the beginning. This I cannot describe, but I saw it, and saw where the Priesthood had been taken from the earth, and how it must be joined together, so that there would be a perfect chain from Father Adam to his latest posterity."[12]

These quotes indicate a spiritual creation in the premortal world. We realize that in the premortal world many entered into covenants with other spirits. Joseph Smith said, "God is good and all his acts [are] for the benefit of inferior intelligences."[13] It is likely that many of the premortal house of Israel, that is, many of us, entered into covenants both with those who would be our ancestors as well as those who would be our posterity. We did this for the express purpose of giving and receiving a saving influence in both directions, progenitors and posterity.

Our own labors with many spirits commenced in the spirit world, our hearts being bound together in love from our associations through eons of immortal existence. There we accepted specific missions to perform a saving work for those with whom we covenanted. Genealogical chains of parents and children were formed, specifically arranged to promote a variety of the Lord's saving purposes. It appears that the plan was that each member of the premortal house of Israel would, as part of his or her progress toward godhood, experience redemption and also learn the role of redeemer; each would be labored with until he or she could labor with others. The redeemed would become the redeemers. To each of us covenant people it would be said, "Freely ye have received, freely give" (Matthew 10:8).

We would, by our premortal covenants with the Lord and our loved ones, become extensions of God's power during our mortal probation, and actually be able to exert a saving influence on an increasing number of people. The Lord teaches Israel about these privileges and responsibilities:

"When men are called unto mine everlasting gospel, and

covenant with an everlasting covenant, they are accounted as the salt of the earth and the savor of men" (D&C 101:39).

"For they [Israel] were set to be a light unto the world, and to be the saviors of men; and inasmuch as they are not the saviors of men, they are as salt that has lost its savor, and is thenceforth good for nothing but to be cast out and trodden under foot of men" (D&C 103:9–10).

As we cease to act in our premortal calling, we ultimately cease to be Israel. The house of Israel, the Lord's covenant people, are the Lord's foreordained army to disseminate the gospel and to save their family connections. Of course, they are "covenant" only insofar as they keep their covenants. That is, the covenant people must continue to choose Christ and promote his purposes and to stand in the office to which they have been called in order to continue as the chosen people. "Chosen" means that they are chosen to a responsibility for which they demonstrated ability and desire in the premortal world. This understanding of the responsibility of the covenant people (see Abraham 2:9: "they shall bear this ministry and Priesthood unto all nations") eliminates any idea of unfair elitism. All who wish may participate and receive the same privileges and assume the same responsibilities. But many people prefer to live according to their own agendas with little interest in what they might do for others (Alma 13:4).

A HOLY CALLING HERE AND HEREAFTER

For Israel, it is a very stirring idea to realize that there are souls depending on us among the dead, the living, and the yet to be born. President Joseph F. Smith taught: "Jesus had not finished his work when his body was slain, neither did he finish it after his resurrection from the dead; although he had accomplished the purpose for which he then came to the earth, he had not fulfilled all his work. And when will he? Not until he has redeemed and saved every son and daughter of our father Adam that have been or ever will be born upon this earth to the end of time, except the sons of perdition. That is his mission. *We will not finish our work until we . . . shall have saved all depending upon us;* for we are to become saviors upon Mount Zion, as well as Christ. We are called to this mission."[14]

This phrase, saving *all souls depending upon us,* that we might become saviors, is truly arresting. It points to the purpose of our lives and at the same time gives us hope that our saving work will continue after this life. Lorenzo Snow taught similarly about our continuing labors with our loved ones after this life:

"God has fulfilled His promises to us, and our prospects are grand and glorious. Yes, in the next life we will have our wives, and our sons and daughters. If we do not get them all at once, we will have them some time, for every knee shall bow and every tongue shall confess that Jesus is the Christ. You that are mourning about your children straying away will have your sons and your daughters. If you succeed in passing through these trials and afflictions and receive a resurrection, you will, by the power of the Priesthood, work and labor, as the Son of God has, until you get all your sons and daughters in the path of exaltation and glory. . . . Therefore, mourn not because all your sons and daughters do not follow in the path that you have marked out to them, or give heed to your counsels. Inasmuch as we succeed in securing eternal glory, and stand as saviors, and as kings and priests to our God, we will save our posterity."[15]

The Prophet Joseph taught that those who keep their temple covenants in this life will have their family in the next: "When a seal is put upon the father and mother, it secures their posterity, so that they cannot be lost, but will be saved by virtue of the covenant of their father and mother."[16]

Elder Boyd K. Packer expanded on this principle of the sealing of the father and mother: "We cannot overemphasize the value of temple marriage, the binding ties of the sealing ordinance, and the standards of worthiness required of them. When parents keep the covenants they have made at the altar of the temple, their children will be forever bound to them."[17] He also quoted Orson F. Whitney concerning the reclaiming of loved ones who stray during this life:

"Though some of the sheep may wander, the eye of the Shepherd is upon them, and sooner or later they will feel the tentacles of Divine Providence reaching out after them and drawing them back to the fold. Either in this life or the life to come, they will return. They will have to pay their debt to justice; they will suffer for their sins; and may tread a thorny path; but if it leads them at last, like the penitent Prodigal, to a loving and forgiving father's heart and home, the painful experience will not have been in vain. Pray for your careless and disobedient children; hold on to them with your faith. Hope on, trust on, till you see the salvation of God."[18]

The desire to hold on to one's children has led many parents to sanctify themselves. Alma describes the nature of sanctification:

"Now, as I said concerning the holy order, or this high priest-hood, there were many who were ordained and became high priests of God; and it was on account of their exceeding faith and repentance, and their righteousness before God, they choosing to repent and work righteousness rather than to perish; therefore they were called after this holy order, and were sanctified, and their garments were washed white through the blood of the Lamb.

"Now they, after being sanctified by the Holy Ghost, having their garments made white, being pure and spotless before God, could not look upon sin save it were with abhorrence; and there were many, exceedingly great many, who were made pure and entered into the rest of the Lord their God" (Alma 13:10–12).

We can see this principle of the parents' sanctification exerting a saving influence in the life of the younger Alma. He had spent many years in this middle world in spiritual darkness before he knew who he was. The account of his redemption from spiritual death really begins with arrangements made before this life, but on earth it may have begun with an incident in his father's life (Mosiah 26) when, as priesthood leader, he responded to a need in the Church. The elder Alma received from King Mosiah the responsibility of judging the rising generation of unbelievers, among whom were the elder Alma's own son and the sons of the king. After pouring out his whole soul to God, fearing that he should do wrong in the sight of God, he received the voice of the Lord:

"Because thou hast inquired of me concerning the transgressor, thou art blessed. Thou art my servant; and I covenant with thee that thou shalt have eternal life; and thou shalt serve me and go forth in my name, and shalt gather together my sheep" (Mosiah 26:19–20).

With the elder Alma's promise of eternal life, his power to draw down grace from his Heavenly Father in behalf of his loved ones increased. His prayers and those of King Mosiah brought an angel to these rebellious sons—most Latter-day Saints know the rest of the story. Later, this redeemed younger Alma exerted this same saving influence over his own erring missionary son, Corianton.

Our father, Joseph of Egypt, set the example for all Israel as he prepared himself to have a sanctifying influence on his troubled family, and like his Savior, exercised a saving power on his brethren. He succeeded in changing family dynamics and in sending a saving influence down many generations. Jesus himself is our model when he says, "For their sakes I sanctify myself" (John 17:19).

One understanding that is essential to this discussion on personal sanctification to save our families is the truth that the people in our families are not there by chance. That fact alerts us to the holy purposes behind having particular people in our family to save. Two presidents of the Church confirmed this truth that people are not placed randomly on the earth.

President Spencer W. Kimball quoted William Law: "It is said that the very hairs of your head are all numbered; is it not to teach us that nothing, not the smallest things imaginable, happen to us by chance? But if the smallest things we can conceive of are declared to be under the divine direction, need we, or can we, be more plainly taught that the greatest things of life, such as the manner of our coming into the world, our parents, the time, and other circumstances of our birth and condition, are all according to the eternal purposes, direction, and appointment of divine Providence?"[19]

And President Harold B. Lee bore this testimony: "You have been blessed to have a physical body because of your obedience to certain commandments in that premortal state. You are now born into a family to which you have come, into the nations through which you have come, as a reward for the kind of lives you lived before you came here and at a time in the world's history, as the Apostle Paul taught the men of Athens [Acts 17:26] and as the Lord revealed to Moses, determined by the faithfulness of each of those who lived before this world was created."[20]

A caution: we want to take care with such a quote as President Lee's that we do not conclude that if one didn't come to a "good" family it was because he or she was not good in the premortal world. That would be a mistaken conclusion. Dr. Carlfred Broderick taught that the Lord may assign a valiant spirit to a troubled family, just as Joseph of Egypt was, in order to bring salvation to that family. We might call this "the Joseph principle":

"Children need not merely replicate the sins of their fathers, . . . each generation is held accountable for its own choices [Ezek. 18:2–4]. Indeed, my experience in various church callings and in my profession as a family therapist has convinced me that God actively intervenes in some destructive lineages, assigning a valiant spirit to break the chain of destructiveness in such families. Although these children may suffer innocently as victims of violence, neglect, and exploitation, through the grace of God some find the strength to 'metabolize' the poison within themselves, refusing to pass it on to future generations. Before them were generations of destructive pain; after them the line flows clear and pure. Their children and children's children will call them blessed.

"In suffering innocently that others might not suffer, such persons, in some degree, become as 'saviors on Mount Zion' by helping to bring salvation to a lineage. . . .

"Others of us may be, ourselves, the suffering messengers of light. Let us be true to our divine commission, forgoing bitterness and following in our Savior's footsteps."[21]

Each of us has the opportunity to be a Joseph in our family. Many who have felt as though they were victims in their families have come to realize that God has actually given us a Joseph role.

Knowing that we are a Joseph in our family helps us to see that we have the opportunity under the Lord's direction "to break the chain of destructiveness" in our own family, refusing to pass onto future generations the toxins of the past. Understanding the "Joseph principle" can lend an entirely different perspective to troubled family relationships.

We see another variation as well, namely, that God may place spiritually challenging children in homes of spiritual and conscientious parents for their mutual benefit. Each has a very specific service to render the other.

Concerning premortal preparations that extend into this life and beyond, President Joseph F. Smith, in his vision of the redemption of the dead, recorded: "I observed that they [leaders of this dispensation] were also among the noble and great ones who were chosen in the beginning to be rulers in the Church of God. Even before they were born, they, with many others, *received their first lessons in the world of spirits and were prepared to come forth* in the due time of the Lord to labor in his vineyard for the salvation of the souls of men.

"I beheld that the faithful elders of this dispensation, *when they depart from mortal life, continue their labors in the preaching of the gospel* of repentance and redemption, through the sacrifice of the Only Begotten Son of God, among those who are in darkness and under the bondage of sin in the great world of the spirits of the dead" (D&C 138:55–57).

An earlier passage teaches that after the righteous are resurrected and enter into the Father's kingdom, they "continue thenceforth their labor as had been promised by the Lord, and [are] partakers of all blessings which were held in reserve for them that love him" (D&C 138:52). It appears, based on this scripture and also on the remarks of President Snow and Elder Hyde, that our work will continue after the resurrection not only for the salvation of our own spirit progeny, but also for our own family members and loved ones from this life and the world before.

When conscientious parents die, their desire for the salvation of their families increases. The Prophet Joseph taught about priesthood holders who have gone on into the spirit world and continue their labors in angelic ministries with their loved ones on earth: "These men are in heaven, but their children are on the earth. Their bowels yearn over us. God sends down men for this reason. 'And the Son of Man shall send forth His angels.' . . . All these authoritative characters will come down and join hand in hand in bringing about this work.

"The Kingdom of Heaven is like a grain of mustard seed. The mustard seed is small, but brings forth a large tree, and the fowls lodge in the branches. The fowls are the angels. Thus angels come down, combine together to gather their children, and gather them. *We cannot be made perfect without them, nor they without us.*"[22]

In another place the Prophet Joseph taught: "The spirits of the just are exalted to a greater and more glorious work; hence they are blessed in their departure to the world of spirits. Enveloped in flaming fire, they are not far from us, and know and understand our thoughts, feelings, and motions, and are often pained therewith."[23]

President Joseph F. Smith, himself a man of deep family feeling, also taught about the feelings of the departed spirits and their nearness to us: "Sometimes the Lord expands our vision from this point of view and this side of the veil, that we feel and seem to realize that we can look beyond the thin veil which separates us from that other sphere. If we can see, by the enlightening influence of the Spirit of God and through the words that have been spoken by the holy prophets of God, beyond the veil that separates us from the spirit world, surely those who have passed beyond, can see more clearly through the veil back here to us than it is possible for us to see to them from our sphere of action.

"I believe we move and have our being in the presence of heavenly messengers and of heavenly beings. We are not separated from them. We begin to realize more and more fully, as we become acquainted with the principles of the gospel, as they have been revealed anew in this dispensation, that we are closely related to our kindred, to our ancestors, to our friends and associates and co-laborers who have preceded us into the spirit world. We cannot forget them;

we do not cease to love them; we always hold them in our hearts, in memory, and thus we are associated and united to them by ties we cannot break. . . .

"[They] can see us better than we can see them— . . . they know us better than we know them. They have advanced; we are advancing; we are growing as they have grown; we are reaching the goal that they have attained unto; and therefore, I claim that we live in their presence, they see us, they are solicitous for our welfare, they love us now more than ever."[24]

Family work is so thoroughly enmeshed in the purposes of the creation of the earth that Moroni quoted Malachi's words to the Prophet Joseph that the whole earth would be wasted if the work of the fathers could not be fulfilled, if fathers and children did not turn to each other. Moroni quoted: "I will reveal unto you the Priesthood, by the hand of Elijah the prophet. . . . And he shall plant in the hearts of the children the promises made to the fathers, and the hearts of the children shall turn to their fathers. If it were not so, the whole earth would be utterly wasted at his coming" (Joseph Smith—History 1:38–39).

Why would the earth be wasted? Because if our hearts did not turn to the premortal promises made among the fathers and children, the fulfillment of which leads us to family sealings, the very purpose for which the earth was created would fail. Family work is the purpose of the earth's creation.

Several years later, the Prophet Joseph reaffirmed the interdependence of family members and the necessity of perfecting family relationships through priesthood power: "The earth will be smitten with a curse unless there is a welding link of some kind or other between the fathers and the children. . . . For we without them cannot be made perfect; neither can they without us be made perfect" (D&C 128:18).

Elder McConkie focused on the meaning of *fathers:* "'He shall plant in the hearts of the children the promises made to the fathers.' That immediately raises the questions: Who are the *children*, who are the *fathers*, and what are the *promises*? If we can catch a vision from the doctrinal standpoint that answers those questions—who the fathers are, who the children are, and what the promises were—we can have our understanding of the gospel and our comprehension of the plan of salvation expanded infinitely. We shall then catch a vision of what the whole system of salvation is all about. Until we do that, really, we never catch that vision."[25]

He identified the fathers as Abraham, Isaac, and Jacob, the fathers of the house of Israel on the earth. The promises have to do with the Abrahamic covenant, which is the premortal covenant of godhood, named after Abraham because he would be one of the fathers of that great lineage.

But perhaps the term *fathers* also refers to those fathers who reach far back into the premortal past where they, preparing to come forth on the earth, said in effect to their Heavenly Father, "I will go down and keep thy commandments and bring these others with me into thy rest."

As we conclude I want to bring this to a personal level. The principle of the premortal house of Israel being arranged in saving relationships helps us to look at ourselves and ask, "In view of the holy purposes for my being, what manner of man or woman do I have the privilege to be?" And having answered that question in spirit, we then may look at all our relationships and ask, "What are my covenant opportunities with respect to the redemption of these people in my life, and how can I fulfill them? What do I need to do now while the window of opportunity is open?" That question, when asked with an honest heart, will bring power and direction from the great Redeemer himself. And not only guidance will come, but also a realization that the window of opportunity will not always be open. There is an urgency to act now to do whatever lies within our power.

The First Presidency's Proclamation on the Family says: "The family is central to the Creator's plan for the eternal destiny of His children."[26] The earth and the very lives we live upon the earth were created for entirely spiritual purposes. Once we know what we're up to here, we can focus on what will really endure. All the activities of men are designed, not as ends in themselves, but as means of getting people together so that they can have a saving influence on each other. When we know that that work is more important than any other, we'll know what it is we are to pay attention to during the minutes and hours of our telestial lives, and at the same time, what we can patiently hope for.

May we take a long-range view, both with our erring loved ones and with ourselves, as we fall so short of all that we wish to be, keeping in our hearts this assurance of Elder Orson F. Whitney: "They have but strayed in ignorance from the Path of Right, and God is merciful to ignorance. Only the fulness of knowledge brings the fulness of accountability. Our Heavenly Father is far more merciful, infinitely more charitable, than even the best of his servants, and the Everlasting Gospel is mightier in power to save than our narrow finite minds can comprehend."[27]

Notes

This is a previously unpublished paper.

1. Joseph Fielding Smith, *Doctrines of Salvation*, 3 vols. (Salt Lake City: Bookcraft, 1954–56), 1:59; emphasis added; see also Deuteronomy 32:7–8.
2. Bruce R. McConkie, address given at BYU Summer Session, 8 August 1967; tape in possession of Robert J. Matthews.
3. Brigham Young, in *Journal of Discourses*, 11:253.
4. *Teachings of the Prophet Joseph Smith*, 365.
5. The emphasis in this verse and in subsequent scriptural passages in this chapter has been added by the author.
6. David B. Haight, in Conference Report, October 1990, 73–74.
7. Joseph Fielding Smith, *The Way to Perfection* (Salt Lake City: Genealogical Society of Utah, 1935), 50–51.
8. Quoted in *Our Lineage* (Salt Lake City: Genealogical Society of Utah, 1933), 4. In this genealogy lesson manual the writer, who is unnamed, makes additional references to premortal preparations just before quoting Elder Woodruff. Because his remarks have interesting specifics and are in harmony with what the Brethren have taught, they are included here: "Many there held important positions of leadership. From the tests of that existence they emerged triumphant. Because of their faithfulness they were accounted worthy to bear in life great responsibilities, and were reserved in training there until a day came in earth's history when the very staunchest and bravest would be needed, as 'tried souls, mid untried spirits found, that captained these may be.' With the dawning of this last Gospel dispensation came their call to journey earthward, and perform the special mission for which they were qualified by character and experience.
9. Spencer W. Kimball, "The Role of Righteous Women," *Ensign*, November 1979, 102.
10. *Teachings of the Prophet Joseph Smith*, 181; emphasis added.
11. Journal History, 23 February 1847.
12. Ibid.; see also Moses 5:10 and Adam's prophecy of all the families of the earth.

13. *Words of Joseph Smith*, 68.
14. *Gospel Doctrine*, 442; emphasis added.
15. *The Teachings of Lorenzo Snow*, ed. Clyde J. Williams (Salt Lake City: Bookcraft, 1998), 195.
16. *Teachings of the Prophet Joseph Smith*, 321; *Words of Joseph Smith*, 241.
17. Boyd K. Packer, Conference Report, April 1992, 94–95.
18. Ibid., 94; see also Brigham Young, in *Journal of Discourses*, 11:215.
19. William Law as quoted in Spencer W. Kimball, "Small Acts of Service," *Ensign*, December 1974, 5.
20. Harold B. Lee, in Conference Report, October 1973, 7.
21. Carlfred Broderick, "I Have A Question," *Ensign*, August 1986, 38–39.
22. *Teachings of the Prophet Joseph Smith*, 159; emphasis added.
23. *Teachings of the Prophet Joseph Smith*, 326.
24. *Gospel Doctrine*, 430–31.
25. Bruce R. McConkie, "Promises Made to the Fathers," in *Studies in Scripture, Genesis to 2 Samuel* (Salt Lake City: Deseret Book Co., 1989), 3:51–52; emphasis in original.
26. "The Family: A Proclamation to the World," *Ensign*, November 1995, 102.
27. Orson F. Whitney, in Conference Report, April 1929, 110.

USING THE BOOK OF MORMON TO FACE THE TESTS AHEAD

Our dispensation will embrace the greatest extremes of any: the tares will manifest greater virulence than in any preceding period. President Ezra Taft Benson has declared: "Wickedness is rapidly expanding in every segment of our society (see D&C 1:14–16; 84:49–53). It is more highly organized, more cleverly disguised, and more powerfully promoted than ever before." But at the same time, the wheat will exhibit greater quality than ever before (D&C 86:1–7), because the Church and kingdom of God are increasing in strength, size, and faithfulness. "It [the Church] has never been better organized or equipped to perform its divine mission."[1] Thus, in the midst of the worst trials ever known among the children of men, the Saints will finish up the work of this dispensation under the direction of the Lord Jesus Christ. It is not a defensive position we wish to take but rather a powerful moving forward to establish all of the will of the Lord in preparation for his coming. We need not shrink before the onslaught but prepare to go out to meet the Bridegroom (D&C 133:10).

We will never meet all the challenges of this dispensation unaided. The forces about to be unleashed against the world will be sufficient to decimate the entire population (D&C 5:19). Only those who have learned how to receive the Holy Spirit for their guide will be able to read the signs and abide those days (D&C 45:57); only those who have studied how to draw on grace, on the divine enabling power of the Lord Jesus Christ, will escape through the means the Lord will provide (D&C 63:34). We will have to be a people who understand personal revelation. We will have to live in direct contact with the temporal world, but we will have to know how to be guided by the heavenly, unseen world.

It is possible to make a long list of specific things we would have to do to endure to the end. But the real question is this: What is that quintessential preparation from which all other preparations naturally follow? The Lord has provided the answer, and it lies in the inspired use of the Book of Mormon—in particular, the use of the Book of Mormon as an instrument of personal revelation. Nearly every description of the tests that the Saints will undergo makes it clear that only personal revelation will make faithful endurance possible. Inspired uses of the Book of Mormon lie at the very core of the Saints' preparation so that they, as the children of light, will not be overtaken as by a thief at the Lord's appearance (D&C 106:4–5).

Prophets through the ages have looked at our day and warned us of the trials that lie ahead. President Heber C. Kimball "often used the language, 'A test, a test is coming.'"[2]

"We think we are secure here in the chambers of the everlasting hills, where we can close those few doors of the canyons against mobs and persecutors, the wicked and the vile, who have always beset us with violence and robbery, but I want to say to you, my brethren, the time is coming when we will be mixed up in these now peaceful valleys to that extent that it will be difficult to tell the face of a Saint from the face of an enemy to the people of God. Then, brethren, look out for the great sieve, for there will be a great sifting time, and many will fall; for I say unto you there is a *test,* a TEST, a TEST coming, and who will be able to stand?"[3]

"This Church has before it many close places through which it will have to pass before the work of God is crowned with victory. To meet the difficulties that are coming, it will be necessary for you to have a knowledge of the truth of this work for yourselves. The difficulties will be of such a character that the man or woman who does not possess this personal knowledge or witness will fall. If you have not got the testimony, live right and call upon the Lord and cease not till you obtain it. If you do not you will not stand. . . .

" . . . The time will come when no man nor woman will be able to endure on borrowed light. Each will have to be guided by the light within himself. If you do not have it, how can you stand?"[4]

President Heber C. Kimball was quoted later with respect to the intensity of the tests: "The judgments of God will be poured out upon the wicked, to the extent that our elders from far and near will be called home; or in other words, the Gospel will be taken from the gentiles, and later on will be carried to the Jews.

"The western boundaries of the State of Missouri will be swept so clean of its inhabitants that as President Young tells us, 'when we return to that place there will not be as much as a yellow dog to wag his tail.'

"Before that day comes, however, the Saints will be put to the test that will try the very best of them.

"The pressure will become so great that the righteous among us will cry unto the Lord day and night until deliverance comes. . . .

"Then is the time to look out for the great sieve, for there will be a great sifting time, and many will fall."[5]

President Ezra Taft Benson expanded on the nature of the tests we will face and confirmed that the process has already begun: "There is a real sifting going on in the Church, and it is going to become more pronounced with the passing of time. It will sift the wheat from the tares, because we face some difficult days, the like of which we have never experienced in our lives. And those days are going to require faith and testimony and family unity, the like of which we have never had."[6]

"The great destructive force which was to be turned loose on the earth and which the prophets for centuries have been calling the 'abomination of desolation' is vividly described by those who saw it in vision (see Matthew 24:15; Joseph Smith—Matthew 1:12, 32). Ours is the first generation to realize how literally these prophecies can be fulfilled now that God, through science, has unlocked the secret to thermonuclear reaction.

"In the light of these prophecies, there should be no doubt in the mind of any priesthood holder that the human family is headed for trouble. There are rugged days ahead. It is time for every man who wishes to do his duty to get himself prepared—physically, spiritually, and psychologically—for the task which may come at any time, as suddenly as the whirlwind."[7]

"We will live in the midst of economic, political, and spiritual instability. When these signs are observed—unmistakable evidences that His coming is nigh—we need not be troubled, but 'stand in holy places, and be not moved, until the day of the Lord come' (D&C 87:8). Holy men and women stand in holy places, and these holy places consist of our temples, our chapels, our homes, and stakes of Zion, which are, as the Lord declares, 'for a defense, and for a refuge from the storm, and from wrath when it shall be poured out without mixture upon the whole earth' (D&C 115:6). We must heed the Lord's counsel to the Saints of this dispensation: 'Prepare yourselves for the great day of the Lord' (D&C 133:10).

"This preparation must consist of more than just casual membership in the Church. *We must be guided by personal revelation and the counsel of the living prophet so we will not be deceived.* Our Lord has indicated who, among Church members, will stand when He appears: 'At that day, when I shall come in my glory, shall the parable be fulfilled which I spake concerning the ten virgins' (D&C 45:56)."[8]

The Lord has given the sobering revelation that in the midst of the latter-day trials, Church members will feel the Lord's refining power as he forges the Saints into a force for fulfilling his divine purposes: "The kingdom of heaven is like unto a net that was cast into the sea, and gathered of every kind, which, when it was full, they drew to shore, and sat down, and gathered the good into vessels; but cast the bad away. So shall it be at the end of the world. And the world is the children of the wicked. The angels shall come forth, and sever the wicked from among the just, and shall cast them out into the world to be burned. There shall be wailing and gnashing of teeth" (JST, Matthew 13:48–51).

"Behold, vengeance cometh speedily upon the inhabitants of the earth, a day of wrath, a day of burning, a day of desolation, of weeping, of mourning, and of lamentation; and as a whirlwind it shall come upon all the face of the earth, saith the Lord. *And upon my house shall it begin, and from my house shall it go forth,* saith the Lord; first among those among you, saith the Lord, who have professed to know my name and have not known me, and have blasphemed against me in the midst of my house, saith the Lord. . . . But purify your hearts before me. . . . Cleanse your hearts and your garments, lest the blood of this generation be required at your hands" (D&C 112:24–26, 28, 33; see also D&C 88:74).[9]

This refining process will prepare a purified Church membership to meet the Lord Jesus Christ. President Benson warned: "It is well that our people understand this principle, so they will not be misled by those apostates within the Church who have not yet repented or been cut off. But there is a cleansing coming. The Lord says that his vengeance shall be poured out 'upon the inhabitants of the earth. . . . And upon my house shall it begin, and from my house shall it go forth, saith the Lord; First among those among you, saith the Lord, who have professed to know my name and have not known me. . . .' (D&C 112:24–26). I look forward to that cleansing; its need within the Church is becoming increasingly apparent."[10]

"Yes, within the Church today there are tares among the wheat and wolves within the flock. As President Clark stated, 'The ravening wolves are amongst us, from our own membership, and they, more than any others, are clothed in sheep's clothing because they wear the habiliments of the priesthood. . . . We should be careful of them.'"[11]

Yet, even as we witness the steady crescendo of sorrows spreading throughout the earth, it is apparent that the Lord has at least a twofold purpose. On the one hand, the wicked will be purged from the Church and from the earth; on the other, the Saints will be purified and refined:

"Behold, the great day of the Lord is at hand; and who can abide the day of his coming, and who can stand when he appeareth? For he is like a refiner's fire, and like fuller's soap; and he shall sit as a refiner and purifier of silver, and *he shall purify the sons of Levi* [the priesthood holders of today; D&C 84:32–34], and purge them as gold and silver, that they may offer unto the Lord an offering in righteousness. Let us, therefore, as a church and a people, and as Latter-day Saints, offer unto the Lord an offering in righteousness" (D&C 128:24).

The present and the future do not hold so many terrors if we believe the Lord's reassuring words that the refining process is under his benevolent control: "Fear not thine enemies, for they are in mine hands and I will do my pleasure with them. My people must be tried in all things, that they may be prepared to receive the glory that I have for them, even the glory of Zion; and he that will not bear chastisement is not worthy of my kingdom" (D&C 136:30–31).

President John Taylor commented on the necessity of the Saints being tested but of the insignificance of the pain of the testing process in relation to the great destiny of the Saints: "It is necessary that we pass through certain ordeals, and that we be tried. But why is it that we should be tried? There is just the same necessity for it now that there was in former times. I heard the Prophet Joseph say, in speaking to the Twelve on one occasion: 'You will have all kinds of trials to pass through. And it is quite as necessary for you to be tried as it was for Abraham and other men of God, and (said he) God will feel after you, and He will take hold of you and wrench your very heart strings, and if you cannot stand it you will not be fit for an inheritance in the Celestial Kingdom of God.' . . .

"But all these personal things amount to but very little. It is the crowns, principalities, the powers, the thrones, the dominions, and the associations with the Gods that we are after, and we are here to prepare ourselves for those things. We are after eternal exaltation in the Celestial Kingdom of God."[12]

In fact, Zion will be built by those who are purified by their sufferings and have learned to endure suffering in order to obey the Lord's every command. Zion will have to suffer in order to be redeemed because her citizens have not learned to obey: "Were it not for the transgressions of my people, speaking concerning the church and not individuals, they might have been redeemed even now.

"But behold, they have not learned to be obedient to the things which I required at their hands, but are full of all manner of evil, and do not impart of their substance, as becometh saints, to the poor and afflicted among them; and are not united according to the union required by the law of the celestial kingdom; and Zion cannot be built up unless it is by the principles of the law of the celestial kingdom; otherwise I cannot receive her unto myself.

"And my people must needs be chastened until they learn obedience, if it must needs be, by the things which they suffer" (D&C 105:2–6).

The sacrificing to obey unlocks the blessings and powers of heaven:

"Verily I say unto you, all among them [the Saints] who know their hearts are honest, and are broken, and their spirits contrite, and are willing to observe their covenants by sacrifice—yea, every sacrifice which I, the Lord, shall command—they are accepted of me. For I, the Lord, will cause them to bring forth as a very fruitful tree which is planted in a goodly land, by a pure stream, that yieldeth much precious fruit" (D&C 97:8–9).

We read in *Lectures on Faith:* "Let us here observe that a religion that does not require the sacrifice of all things never has power sufficient to produce the faith necessary unto life and salvation. . . . It is through the medium of the sacrifice of all earthly things that men do actually know that they are doing the things that are well pleasing in the sight of God. . . .

" . . . And in the last days, before the Lord comes, he is to gather together his saints who have made a covenant with him by sacrifice."[13]

Elder John Taylor encouraged the Saints to cleave to the Lord as the night darkens: "In relation to events that will yet take place, and the kind of trials, troubles, and sufferings which we shall have to cope with, it is to me a matter of very little moment; these things are in the hands of God, he dictates the affairs of the human family, and directs and controls our affairs; and the great thing that we, as a people, have to do is to seek after and cleave unto our God, to be in close affinity with him, and to seek for his guidance, and his blessing and Holy Spirit to lead and guide us in the right path. Then it matters not what it is nor who it is that we have to contend with, God will give us strength according to our day."[14]

Another source of reassurance is the knowledge that our coming to earth at this time in the earth's history was no random event. We were prepared before we came to earth to do the work we would be called to do (D&C 138:56). President Ezra Taft Benson taught:

"For nearly six thousand years, God has held you in reserve to make your appearance in the final days before the second coming of the Lord. Some individuals will fall away; but the kingdom of God will remain intact to welcome the return of its head—even Jesus Christ. While our generation will be comparable in wickedness to the days of Noah, when the Lord cleansed the earth by flood, there is a major difference this time. It is that God has saved for the final inning some of His strongest children, who will help bear off the kingdom triumphantly. That is where you come in, for you are the generation that must be prepared to meet your God. . . .

" . . . Make no mistake about it—you are a marked generation. There has never been more expected of the faithful in such a short period of time than there is of us. Never before on the face of this earth have the forces of evil and the forces of good been as well organized. Now is the great day of the devil's power. But now is also the great day of the Lord's power, with the greatest number ever of priesthood holders on the earth."[15]

Having faith in our premortal preparation, we can find unique blessings in realizing that the personal tests we face were specially suited to our individual spiritual needs. Perhaps the greatest insight to come out of this realization is that the Lord orchestrates the details in the lives of his seeking and obedient Saints. With his orchestration of their individual trials comes also deliverance from those trials.

Elder Bruce R. McConkie listed the tasks that lie ahead of these premortally prepared Saints, tasks that will require a greater commitment to spiritual principles than ever before: "We have yet to gain that full knowledge and understanding of the doctrines of salvation and the mysteries of the kingdom that were possessed by many of the ancient Saints. O that we knew what Enoch and his people knew! Or that we had the sealed portion of the Book of Mormon, as did certain of the Jaredites and Nephites! How can we ever gain these added truths until we believe in full what the Lord has already given us in the Book of Mormon, in the Doctrine and Covenants, and in the inspired changes made by Joseph Smith in the Bible? . . .

"We have yet to attain that degree of obedience and personal righteousness which will give us faith like the ancients: faith to multiply miracles, move mountains, and put at defiance the armies of nations; faith to quench the violence of fire, divide seas and stop the mouths of lions; faith to break every band and to stand in the presence of God. Faith comes in degrees. Until we gain faith to heal the sick, how can we ever expect to move mountains and divide seas?

"We have yet to receive such an outpouring of the Spirit of the Lord in our lives that we shall all see eye to eye in all things, that every man will esteem his brother as himself, that there will be no poor among us. . . . As long as we disagree as to the simple and easy doctrines of salvation, how can we ever have unity on the complex and endless truths yet to be revealed?

"We have yet to perfect our souls, by obedience to the laws and ordinances of the gospel, and to walk in the light as God is in the light, so that if this were a day of translation we would be prepared to join Enoch and his city in heavenly realms. How many among us are now prepared to entertain angels, to see the face of the Lord, to go where God and Christ are and be like them? . . .

"We have yet to prepare a people for the Second Coming. . . .

" . . . Shall we not now, as a Church and as a people and as the Saints of latter days, build on the foundations of the past and go forward in gospel glory until the great Jehovah shall say: 'The work is done; come ye, enter the joy of your Lord; sit down with me on my throne; thou art now one with me and my Father.'"[16]

On another occasion, Elder McConkie made a sobering statement on the necessity of the development of our faith: "It may be, for instance, that nothing except the power of faith and the authority of the priesthood can save individuals and congregations from the atomic holocausts that surely shall be."[17]

President Brigham Young also observed that, in general, the Saints are not prepared for the blessings that the Lord anticipates bestowing on them: "Jesus has been upon the earth a great many more times than you are aware of. When Jesus makes his next appearance upon the earth, but few of this Church and kingdom will be prepared to receive him and see him face to face and converse with him; but he will come to his temple. . . .

"When all nations are so subdued to Jesus that every knee shall bow and every tongue shall confess, there will still be millions on the earth who will not believe in him; but they will be obliged to acknowledge his kingly government. You may call that government ecclesiastical, or by whatever term you please; yet there is no true government on earth but the government of God, or the holy Priesthood. Shall I tell you what that is? In short, it is a perfect system of government—a kingdom of Gods and angels and all beings who will submit themselves to that government. There is no other true government in heaven or upon the earth. . . .

"Is man prepared to receive that government? He is not. I can say to these Latter-day Saints, You are not prepared to receive that government. You hear men and women talk about living and abiding a celestial law, when they do not so much as know what it is, and are not prepared to receive it. We have a little here and a little there given to us, to prove whether we will abide that portion of law that will enable us to enjoy a resurrection with the just. . . .

" . . . We have line upon line, precept upon precept, here a little and there a little, and it is something that accords with the capacity of finite beings, and you improve upon this, and the Lord will open your minds to receive more, and let you see the order of the eternal

Priesthood; *but if you do not live your religion, you cannot receive more.*"[18]

Many of us are not yet living our religion. The Saints can reach for more. In the foregoing statements of the Lord and his prophets and apostles, one common observation is that amidst the latter-day trials, the Saints will need familiarity with the Lord's voice. For example, President Heber C. Kimball said, "Each will have to be guided by the light within himself. If you do not have it, how can you stand?"[19] President Ezra Taft Benson remarked on the current state of the Saints' preparedness to meet the Savior: "Watchmen—what of the night? We must respond by saying that all is not well in Zion. As Moroni counseled, we must *cleanse the inner vessel* (see Alma 60:23), beginning first with ourselves, then with our families, and finally with the Church. . . . It takes a Zion people to make a Zion society, and *we must prepare for that.*"[20]

Obviously our responses to the Lord's direction have as yet been inadequate to achieve that state of preparation, even though the Lord has made it abundantly clear that the instrument of preparation for the Second Coming is the Book of Mormon. That book is the instrument by which the citizens of Zion will have cleansed the inner vessel.

The Saints, evidently, are still under condemnation for their neglect of the Book of Mormon, the very tool that has the most power to prepare the Church for the advent of the Savior. Upon reflection, we realize that the Saints have not yet adequately made the connection between the comprehensive use of the Book of Mormon and the light that each of them will need to withstand the trials of the latter days. President Benson exclaimed: "Now we not only need to *say* more about the Book of Mormon, but we need to *do* more with it. Why? The Lord answers: 'That they may bring forth fruit meet for their Father's kingdom; otherwise there remaineth a scourge and judgment to be poured out upon the children of Zion' [D&C 84:58]. We have felt that scourge and judgment!"[21]

Perhaps some have thought that a testimony and a general knowledge of the Book of Mormon were sufficient fulfillment of the Lord's injunction. But a testimony of the Book of Mormon is not an end in itself. It is only the most rudimentary beginning. The next step, after learning that the book is true and can be trusted as a source for doctrine and the Spirit, is to learn its multiple uses and virtues. Here is the Lord's fuller text to the Church about using the Book of Mormon:

"Your minds in times past have been darkened because of unbelief, and because you have *treated lightly the things you have received*—which vanity and unbelief have brought the *whole church* under condemnation. And this condemnation resteth upon the children of Zion, *even all*.

"And they shall remain under this condemnation until they repent and remember the new covenant, *even the Book of Mormon* and the former commandments which I have given them, *not only to say, but to do according to that which I have written*—that they may bring forth fruit meet for their Father's kingdom; otherwise there remaineth a scourge and judgment to pour out upon the children of Zion" (D&C 84:54–58).

The Church apparently does not yet know all the uses and virtues of the Book of Mormon. It seems that the Lord would like us to use the Book of Mormon in ways it has not been used before and that he is waiting for us to ask his help to that end. President Benson, in urging us to get more deeply into the book, pointed to the relationship between scripture study and the power of the Spirit in our life: "I urge you to recommit yourselves to a study of the scriptures. Immerse yourselves in them daily *so you will have the power of the Spirit to attend you* in your callings. Read them in your families and teach your children to love and treasure them. Then prayerfully, and in counsel with others, seek every way possible to encourage the members of the Church to follow your example. If you do so, you will find, as Alma did, that 'the word [has] a great tendency to lead people to do that which [is] just—yea, it [has] more powerful effect upon the minds of the people than the sword, or anything else, which [has] happened unto them.' (Alma 31:5.)"[22]

In that statement, President Benson has connected immersion in the scripture with the gift and power of the Spirit. Several other

Brethren have likewise pointed out the link between personal revelation and a spiritually skilled use of scripture. For example, Elder
Bruce R. McConkie said: "I sometimes think that one of the best-
kept secrets of the kingdom is that the scriptures open the door to the
receipt of revelation."[23] "However talented men may be in adminis-
trative matters; however eloquent they may be in expressing their
views; however learned they may be in the worldly things—they will
be denied the sweet whisperings of the Spirit that might have been
theirs unless they pay the price of studying, pondering, and praying
about the scriptures."[24]

Elder Dallin H. Oaks declared: "As a source of knowledge, the
scriptures are not the *ultimate* but the penultimate. *The ultimate
knowledge comes by revelation. . . .* A study of the scriptures enables
men and women to receive revelations. . . . because scripture reading puts us in tune with the Spirit of the Lord."[25]

Elder Boyd K. Packer taught: "Buildings and budgets, and
reports and programs and procedures are very important. But, by
themselves, they do not carry that essential spiritual nourishment and
will not accomplish what the Lord has given us to do. . . . The right
things, those with true spiritual nourishment, are centered in the
scriptures."[26]

One reason that some Church members are not reaping the full
reward of scripture study may be that they do not know how vital the
reward could be. The Lord promises that we can hear or feel his
voice in the scriptures, that we can receive messages in and above
what is printed on the page, and that we can repeat that experience
over and over again. Scripture reading and feasting on scripture can
take on new meaning.

Those who haven't heard the voice may find the scriptures less
interesting than other literature. Perhaps some are afraid to hear
what the Lord has to say, and so they may read scripture with a pro-
tective veil over their minds and then say they are bored with the
scripture. The real problem may be that they are afraid to hear the
voice. But the Lord says to us today, "Resist no more my voice"
(D&C 108:2).

It appears that the Lord intends an intersection between the daily orbit we move in and the orbit of the Book of Mormon. A power is set in motion when we become deeply involved in what the scriptures are saying to us personally. The Lord has said that "the Book of Mormon and the holy scriptures are given of me for your instruction; and the power of my Spirit quickeneth all things" (D&C 33:16).

Several other scriptures also show that the printed word can yield the living spirit of prophecy and revelation to the alert and prepared reader. In the very first chapter of the Book of Mormon, Lehi learned this connection between feasting on scripture and receiving the power of revelation: "As he read [a book of scripture given him by the Lord], he was filled with the Spirit of the Lord" (1 Nephi 1:12). Nephi said, "And now when my father saw all these things [scriptures on the plates of brass], he was filled with the Spirit, and began to prophesy" (1 Nephi 5:17). Jacob made the same connection: "Wherefore, we search the prophets, and we have many revelations and the spirit of prophecy; and having all these witnesses we obtain a hope, and our faith becometh unshaken, insomuch that we truly can command in the name of Jesus and the very trees obey us, or the mountains, or the waves of the sea" (Jacob 4:6). The four sons of Mosiah "waxed strong in the knowledge of the truth; for they were men of a sound understanding and they had searched the scriptures diligently, that they might know the word of God. But this is not all; they had given themselves to much prayer, and fasting; therefore they had the spirit of prophecy, and the spirit of revelation, and when they taught, they taught with power and authority of God" (Alma 17:2–3).

The Lord, speaking of scripture, said: "These words are not of men nor of man, but of me; . . . for it is my voice which speaketh them unto you; for they are given by my Spirit unto you . . . ; wherefore, you can testify that you have heard my voice, and know my words" (D&C 18:34–36). To hear the voice of the Lord in scripture simply means to *feel* the Spirit of the Lord, because the Lord speaks "by the voice of my Spirit" (D&C 75:1). Furthermore, the Lord's "voice is Spirit" (D&C 88:66).

Thus there is a relationship between the written scripture and the voice of the Lord, or personal revelation. The same Spirit that gave the written word quickens it as one who is prepared reads it. Taking all these insights together, we may conclude that if we wish to guide our life by the Spirit, we cannot do it without also being a spiritual student of the living Book of Mormon.

The Book of Mormon itself teaches the progression from feasting on the word of Christ to hearing the voice of Christ through the Holy Ghost: "Feast upon the words of Christ; for behold, the words of Christ will *tell* you all things what ye should do. . . . [I]f ye will enter in by the way, and receive the Holy Ghost, it will *show* unto you all things what ye should do" (2 Nephi 32:3, 5).

The stories and principles and doctrines in the Book of Mormon are vitally important to the Latter-day Saints, but we soon discover that an important principle taught by the book is that no collection of writings can tell a person what to do in all circumstances. Many of life's challenges are designed to require divine insight and divine power and divine direction to meet them. Therefore, perhaps no principle is stressed as much as getting the Spirit of the Lord, who "will show unto you all things what ye should do," as a constant guide.

We can see how the voice of the Lord is a vital component of the gospel plan—not just for prophets but for all of us. Ultimately, everyone who hopes to see the face of the Lord and to remain in his presence must learn to discern and obey the voice of the Lord. That skill is essential for the serious candidate for exaltation.

Elder Richard G. Scott explicitly stated that the Book of Mormon is like a personal Liahona or Urim and Thummim: "What does the Book of Mormon mean to you? . . .

"If you have not yet drunk deeply from this fountain of pure truth, with all of my soul I encourage you to do so now. Don't let the consistent study of the Book of Mormon be one of the things that you intend to do but never quite accomplish. Begin today.

"I bear witness that *it can become a personal 'Urim and Thummim' in your life*."[27]

Indeed, the Book of Mormon seems to describe itself as a

Liahona (Alma 37:44). A primary use of a Liahona or a Urim and Thummim is as a physical symbol to teach the dynamics of revelation. Such objects increase faith until one has learned to get revelation without sole dependence on the physical instrument. How important it is to realize that scripture, as another form of Liahona or Urim and Thummim, sensitizes and instructs our spirits in the processes of revelation.

The link between a tangible object of revelation and the process of receiving revelation without an instrument is illustrated in the Lord's training of the Prophet Joseph Smith. The Lord started Joseph out with the Urim and Thummim; later, Joseph was able to receive revelation without using it, thus showing that a Liahona and Urim and Thummim, seer stones, and scripture are all variations of the sacred instruments by which a person is taught how to receive increasingly detailed revelation—revelation that is often outside the imagination or experience of the person being so trained. Joseph's experience in translating the Book of Mormon by the Urim and Thummim actually prepared him to be the founding prophet, seer, and revelator of this dispensation.

Alma understood this relationship between instruments of revelation and scripture. In Alma 37, Alma used a succession of words that suggests that relationship: *records, plates of brass, holy scriptures, mysteries, holy writ, interpreters, Gazelem, stone* (as in seer stone)*, counsel with the Lord,* and *Liahona* (vv. 1–5, 20, 23, 37–38). When the Nephites used their Liahona, they had miracles every day; whenever they grew lazy and forgot to exercise their faith and diligence, they lost their way and became hungry.

Alma taught that the ball, director, compass, or Liahona was prepared by the Lord as a type of the word of Christ: "For behold, it is as easy to give heed to the word of Christ, which will point to you a straight course to eternal bliss, as it was for our fathers to give heed to this compass, which would point unto them a straight course to the promised land. And now I say, is there not a type in this thing? For just as surely as this director did bring our fathers, by following its course, to the promised land, shall the words of Christ, if we follow their course, carry us beyond this vale of sorrow into a far better land of promise. O my son, do not let us be slothful because of the easiness of the way" (Alma 37:44–46). It is possible that "vale of sorrow" means not only this mortal life as a whole but the individual vales of sorrow the Saints come upon in their lives. That is, the Lord has given the Saints a Liahona to carry them out of their individual and collective vales of sorrow.

One problem of mortality is the inadequacy of our present language to describe spiritual experience. When missionaries try to teach investigators what the Spirit is, they have to do it by analogy or by metaphor; thus, when they see the Spirit working on an investigator, they will say, "That's it! What you are feeling right now is the Spirit of the Lord!" Elder Boyd K. Packer taught: "We do not have the words (even the scriptures do not have words) which perfectly describe the Spirit. The scriptures generally use the word voice, which does not exactly fit. These delicate, refined spiritual communications are not seen with our eyes, nor heard with our ears. And even though it is described as a voice, it is a voice that one feels, more than one hears."[28]

We identify the Spirit mostly by feeling, and through our involvement with the Spirit, which quickens the scriptures, the Lord teaches us what the Spirit feels like. If we think we have to feel something extraordinary in hearing the voice of the Lord in the scriptures, we might miss the subtle impressions of the Spirit. Many people have experienced the movement of the Spirit in their souls as they read scripture. On some occasions, feelings come, or maybe tears, perhaps heightened appreciation, or a sense of peace on a particular matter, a sense of unexplainable happiness, or a sense of the Lord's love. If we are asked on such an occasion what we heard from the Lord in that experience with scripture, we might not be able to articulate an answer. Nonetheless, we have felt something sweet, something very tender. That was the spirit of prophecy and revelation, the Spirit of the Lord Jesus Christ.

On other occasions, we may be reading along when an issue or problem in our life comes to mind, maybe even a subject unrelated to the scripture being read, and suddenly we just know what to do about it. All these are instances of feeling the voice of the Lord speaking to us. We could develop this skill to a high degree and enjoy a living relationship with the Lord, in which the Lord could teach us many wonderful things. To develop such a skill we must invest time to gain experience, must labor in the Spirit, and must make scripture study a part of our daily life, but it is within the capability of any serious seeker.

The Book of Mormon teaches us how to receive revelation from scripture: "I, Nephi, beheld the pointers which were in the ball, that they did work according to the faith and diligence and heed which we did give unto them. And there was also written upon them a new writing, which was plain to be read, which did give us understanding concerning the ways of the Lord; and it was written and changed from time to time, according to the faith and diligence which we gave unto it. And thus we see that by small means the Lord can bring about great things" (1 Nephi 16:28–29).

By studying the spiritual conditions under which the Book of Mormon was translated, we can learn more about the process of receiving the Spirit from scripture. Joseph Smith showed the way. David Whitmer described what the Prophet had to go through to get the spirit of prophecy so that he could translate:

"At times when brother Joseph would attempt to translate . . . , he found he was spiritually blind and could not translate. He told us that his mind dwelt too much on earthly things, and various causes would make him incapable of proceeding with the translation. When in this condition he would go out and pray, and when he became sufficiently humble before God, he could then proceed with the translation. Now we see how very strict the Lord is, and how he requires the heart of man to be just right in his sight before he can receive revelation from him."[29]

On another occasion David Whitmer recorded: "He [Joseph Smith] was a religious and straightforward man. . . . He had to trust in God. He could not translate unless he was humble and possessed the right feelings towards everyone. To illustrate so you can see: One morning when he was getting ready to continue the translation, something went wrong about the house and he was put out about it. Something that Emma, his wife, had done. Oliver and I went upstairs and Joseph came up soon after to continue the translation but he could not do anything. He could not translate a single syllable. He went downstairs, out into the orchard, and made supplication to the Lord; was gone about an hour—came back to the house, and asked Emma's forgiveness and then came upstairs where we were and then the translation went on all right. He could do nothing save he was humble and faithful."[30]

This account is highly instructive. We must approach scripture in the same way that the scripture was given to one who was in a state of humility, of desire, of courage, of forgiveness of others.

We can prepare ourselves to hear the word of the Lord by realizing that in opening up the scriptures, we are about to have a conversation with the Lord. We might say deep in our spirit, "Lord, what dost Thou wish to say to me today?" Thus we approach such an encounter in a spiritual, prayerful, thoughtful, and solemn way. We read trying to feel, to listen, to hear, and even to make notes. Our heart must be prepared to be written on; we must want to hear what the Lord wants to say to us, what the Lord's counsel is to us. So we approach scripture with as much humility as we can, with willingness to repent and to grow.

Whenever we feel that movement of the Spirit in our own soul, we are feeling the voice of the Lord to us—the Lord is speaking to us individually. *We are connected by the Spirit in that moment to our Savior.* In this way, the Book of Mormon can bring us to Christ every time we pick up the book and hear or feel the Spirit. Coming to Christ is the main objective of all scripture. Simply, we feast on the words of Christ and, if we have prepared, he speaks to us through feelings and impressions and happiness and even words, and thus we literally come to Christ as we study scripture and hear his voice.

This kind of immersion in reading, this knowledge of the Book of Mormon, this learning to discern, to hear or feel, and then to obey the voice of the Lord to us personally may do more to prepare the

Saints for the coming of the Lord Jesus Christ than nearly any other activity we could engage in. President Benson urged the priesthood holders of the Church: "One of the most important things you can do as priesthood leaders is to immerse yourselves in the scriptures. Search them diligently. Feast upon the words of Christ. Learn the doctrine. Master the principles that are found therein. There are few other efforts that will bring greater dividends to your calling. There are few other ways to gain greater inspiration as you serve.

"But that alone, as valuable as it is, is not enough. You must also bend your efforts and your activities to stimulating meaningful scripture study among the members of the Church. Often we spend great effort in trying to increase the activity levels in our stakes. We work diligently to raise the percentages of those attending sacrament meetings. We labor to get a higher percentage of our young men on missions. We strive to improve the numbers of those marrying in the temple. All of these are commendable efforts and important to the growth of the kingdom. But when individual members and families immerse themselves in the scriptures regularly and consistently, these other areas of activity will automatically come. Testimonies will increase. Commitment will be strengthened. Families will be fortified. Personal revelation will flow. . . .

" 'This book of the law shall not depart out of thy mouth; but thou shalt meditate therein day and night, that thou mayest observe to do according to all that is written therein: for *then thou shalt make thy way prosperous, and then thou shalt have good success.'* (Josh. 1:8; italics added)."[31]

President Benson has also spoken on the centrality of the scriptures to the work that the Saints must do in the winding-up scenes of this dispensation: "In the Book of Mormon we find a pattern for preparing for the Second Coming. A major portion of the book centers on the few decades just prior to Christ's coming to America. By careful study of that time period we can determine why some were destroyed in the terrible judgments that preceded His coming and what brought others to stand at the temple in the land of

Bountiful and thrust their hands into the wounds of His hands and feet."[32]

"My beloved brothers and sisters, I bear my solemn witness that these books [the Book of Mormon and the Doctrine and Covenants] contain the mind and the will of the Lord for us in these days of trial and tribulation. They stand with the Bible to give witness of the Lord and His work. These books contain the voice of the Lord to us in these latter days. *May we turn to them with full purpose of heart and use them in the way the Lord wishes them to be used.*"[33]

It is clear that as wickedness increases, the Saints need a compensatory blessing to carry the Lord's work forward. We have been promised that very blessing. President Benson declared: "I bless you with increased *understanding* of the Book of Mormon. I promise you that from this moment forward, if we will daily sup from its pages and abide by its precepts, God will pour out upon each child of Zion and the Church a blessing hitherto unknown."[34]

Finally, President Benson has spoken on how we will get from where we are now to that day when our Savior appears and the Saints stand before him prepared: "Only a Zion people can bring in a Zion society. And as the Zion people increase, so we will be able to incorporate more of the principles of Zion until we have a people prepared to receive the Lord."[35]

When the Saints have assumed their individual responsibility to possess the Spirit of the Lord in the ways that the Lord has instructed, all other preparations will follow, and we will have a people not only ready to stand in the midst of the trials preceding the Second Coming but able to rejoice under the sanctifying and prospering hand of the Lord Jesus Christ.

Notes

Revised from *Watch and Be Ready: Preparing for the Second Coming of the Lord* (Salt Lake City: Deseret Book Co., 1994), 16–39.

1. Ezra Taft Benson, in Conference Report, October 1988, 103.
2. Orson F. Whitney, *Life of Heber C. Kimball*, 2d ed. (Salt Lake City: Stevens and Wallis, 1945), 447.
3. Ibid., 446; emphasis in original.

4. Ibid., 449–50.
5. J. Golden Kimball, in Conference Report, October 1930, 59–60.
6. Ezra Taft Benson, *The Teachings of Ezra Taft Benson* (Salt Lake City: Bookcraft, 1988), 107.
7. Benson, *Teachings of Ezra Taft Benson,* 107–8.
8. Benson, *Teachings of Ezra Taft Benson,* 106–7; emphasis added.
9. The emphasis in this verse, and in subsequent scriptural passages in this chapter, has been added by the author.
10. Ezra Taft Benson, in Conference Report, April 1969, 10.
11. Benson, in Conference Report, April 1969, 11, citing J. Reuben Clark Jr., in Conference Report, April 1949, 153.
12. John Taylor, in *Journal of Discourses,* 24:197–98.
13. *Lectures on Faith,* 69–70.
14. John Taylor, in *Journal of Discourses,* 18:281.
15. Benson, *Teachings of Ezra Taft Benson,* 104–5.
16. Bruce R. McConkie, "This Final Glorious Gospel Dispensation," *Ensign,* April 1980, 25.
17. Bruce R. McConkie, "Stand Independent above All Other Creatures," *Ensign,* May 1979, 93.
18. Brigham Young, in *Journal of Discourses,* 7:142–43; emphasis added.
19. Whitney, *Heber C. Kimball,* 450.
20. Ezra Taft Benson, "Cleansing the Inner Vessel," *Ensign,* May 1986, 4; emphasis added.
21. Benson, "Cleansing the Inner Vessel," 5; emphasis in original.
22. Benson, "The Power of the Word," *Ensign,* May 1986, 82; emphasis added.
23. Bruce R. McConkie, *Doctrines of the Restoration: Sermons and Writings of Bruce R. McConkie,* ed. Mark L. McConkie (Salt Lake City: Bookcraft, 1989), 243.
24. Bruce R. McConkie, quoted in Benson, "The Power of the Word," 81.
25. Dallin H. Oaks, "Scripture Reading and Revelation," address delivered to BYU Studies Academy, Provo, Utah, 29 January 1993, 3–4; emphasis in original.
26. Boyd K. Packer, quoted in Benson, "The Power of the Word," 81.
27. Richard G. Scott, "The Power of the Book of Mormon in My Life," *Ensign,* October 1984, 11; emphasis added.
28. Boyd K. Packer, "The Candle of the Lord," *Ensign,* January 1983, 52.
29. David Whitmer, *Address to All Believers in Christ;* in B. H. Roberts, *A Comprehensive History of The Church of Jesus Christ of Latter-day Saints, Century One,* 6 vols. (Provo, Utah: The Church of Jesus Christ of Latter-day Saints, 1957), 6:130–31.
30. David Whitmer, in Roberts, *Comprehensive History,* 1:131.
31. Benson, "The Power of the Word," 81; emphasis in original.
32. Benson, *Teachings of Ezra Taft Benson,* 59.
33. Ezra Taft Benson, "The Gift of Modern Revelation," *Ensign,* November 1986, 80; emphasis added.
34. Benson, *Ensign, "A Sacred Responsibility,"* May 1986, 78; emphasis in original.
35. Ezra Taft Benson, "Jesus Christ—Gifts and Expectations," *Speeches of the Year, 1974* (Provo, Utah: Brigham Young University Press, 1975), 305.

Chapter 4

TYPES AND SHADOWS OF DELIVERANCE IN THE BOOK OF MORMON

Grasping the Lord's outstretched hand for help requires reaching into the unknown for the unseen. To assist the humble seeker of Christ to bridge that gap, the Lord provided the Book of Mormon. This book presents a series of dilemmas that are types of the troubles that men and women face in all dispensations: being lost, hungry, enslaved, in danger, or possessed by such painful emotions as anger, guilt, depression, and fear—situations from which people need deliverance.

Deliverance from such trouble is a major theme of the Book of Mormon. A computer count shows that the words derived from *deliver* occur more than two hundred times in the 531 pages of the Book of Mormon, signifying the importance of the principle.[1] The reader repeatedly learns that God will provide some deliverance from trouble if he will but turn to Him. The Book of Mormon speaks to all ages, and its principles apply to all people everywhere. No one can ever have a dilemma that the Lord cannot turn into some form of deliverance. The purpose of this chapter is not only to heighten our sensitivity to the concept of deliverance in the Book of Mormon, and thereby increase our faith in the accessibility of Christ's help, but also to point out the principles by which deliverance is obtained.

During his mortal probation, man is in a bondage that he may not even perceive. The Book of Mormon seeks to teach man that he has an acute need for deliverance from the bondages of his fallen condition. Even though many different kinds of deliverance are described, the ultimate object of all the deliverances is to bring that which is miserable, scattered, alienated, and spiritually dead back into living oneness with Christ: deliverance is a function of the power of at-one-ment in Jesus Christ. Jacob explained:

"And because of the way of *deliverance* of our God, the Holy One of Israel, this death, of which I have spoken, which is the temporal, shall *deliver* up its dead; . . . which spiritual death is hell. . . . O the greatness of the mercy of our God, the Holy One of Israel! for he *delivereth* his saints from that awful monster the devil, and death, and hell, and that lake of fire and brimstone, which is endless torment" (2 Nephi 9:1–12, 19).[2] Book of Mormon accounts of deliverance point the reader's mind to the greatest deliverance of all, the redemption of mankind from physical and spiritual death by the Lord Jesus Christ.

We find the theme of deliverance in the first chapter of the Book of Mormon, signifying its preeminence: "Behold, I, Nephi, will show unto you that the tender mercies of the Lord are over all those whom he hath chosen, because of their faith, to make them mighty even unto the power of *deliverance*" (1 Nephi 1:20).

Clearly *deliverance* is a key word selected by Nephi, under the Lord's inspiration, to set an important theme of the entire Book of Mormon. Following are some random samples of the use of *deliverance* in the Book of Mormon:

Nephi explained to his fearful brothers, "The Lord is able to deliver us, even as our fathers, and to destroy Laban, even as the Egyptians" (1 Nephi 4:3).

Alma rebuked the unbelief of the people of Ammonihah: "Have ye forgotten so soon how many times he *delivered* our fathers out of the hands of their enemies, and preserved them from being destroyed, even by the hands of their own brethren?" (Alma 9:10).

And again: "I would that ye should do as I have done, in remembering the captivity of our fathers; for they were in bondage, and none could *deliver* them except it was the God of Abraham, and the

God of Isaac, and the God of Jacob; and he surely did *deliver them in their afflictions*" (Alma 36:2).

Helaman wrote of his experiences with his two thousand stripling warriors: "We did pour out our souls in prayer to God, that he would strengthen us and *deliver* us out of the hands of our enemies. . . . Yea, and it came to pass that the Lord our God did visit us with assurances that he would *deliver* us; yea, insomuch that he did speak peace to our souls, and did grant unto us great faith, and did cause us that we should hope for our *deliverance* in him" (Alma 58:10–11).

A systematic survey of the fifteen books of the Book of Mormon reveals how well the idea of deliverance is spread through its pages. Such a wide distribution demonstrates that Nephi, Jacob, and Mormon used *deliverance* as one of the organizing principles of the Book of Mormon. Following are examples of its distribution:

First Nephi. "And I, Nephi, beheld that the Gentiles that had gone out of captivity were *delivered* by the power of God out of the hands of all other nations" (1 Nephi 13:19).

Second Nephi. "O house of Israel, is my hand shortened at all that it cannot redeem, or have I no power to *deliver?*" (2 Nephi 7:2).

Jacob. The prophet Jacob teaches the concept of deliverance but does not use the word in his own book. However, the word does appear in 2 Nephi 6:17 (from Isaiah 49: "For the Mighty God shall deliver his covenant people") and 2 Nephi 9:10–13, both of which are Jacob's writings.

Enos and Jarom. Neither Enos's twenty-seven verses nor Jarom's fifteen verses include this sense of the word *deliver*. Nevertheless, Enos 1:15 conveys the concept of deliverance: "Whatsoever thing ye shall ask in faith, believing that ye shall receive in the name of Christ, ye shall receive it."

Omni. "Wherefore, the Lord did visit them in great judgment; nevertheless, he did spare the righteous that they should not perish, but did *deliver* them out of the hands of their enemies" (Omni 1:7).

Words of Mormon. The eighteen verses of Words of Mormon teach the concept but do not use the word *deliver* in our sense.

Mosiah. "Put your trust in him, and serve him with all diligence of mind, [and] if ye do this, he will, according to his own will and pleasure, *deliver* you out of bondage" (Mosiah 7:33).

Alma. "God would make it known unto them whither they should go to defend themselves against their enemies, and by so doing, the Lord would *deliver* them; and this was the faith of Moroni, and his heart did glory in it" (Alma 48:16).

Helaman. "O, how could you have forgotten your God in the very day that he has *delivered* you?" (Helaman 7:20).

Third Nephi. "As the Lord liveth, except ye repent of all your iniquities, and cry unto the Lord, ye will in no wise be *delivered* out of the hands of those Gadianton robbers" (3 Nephi 3:15).

Fourth Nephi. The forty-nine verses of 4 Nephi do not use the term *deliverance*.

Mormon. The Lord said: "And thrice have I *delivered* them out of the hands of their enemies, and they have repented not of their sins" (Mormon 3:13).

Ether. The book of Ether expresses the concept of deliverance: for example, "Therefore when they were encompassed about by many waters they did cry unto the Lord, and he did bring them forth again upon the top of the waters" (Ether 6:7).

The Book of Mormon provides many examples and types of deliverance that range from saving an entire nation, as in the often-evoked story of the exodus of the children of Israel out of Egypt,[3] to the individual deliverance for which Nephi pleaded (2 Nephi 4:27–33). An important point here is that the Lord has provided with these examples of deliverance instructions for obtaining deliverance for oneself and one's community.

For example, nearly every person has experienced the pain and confusion of spiritual darkness. Helaman 5 offers a formula for dispersing spiritual and emotional darkness as it describes the fearful cloud that descended upon the Lamanites who had imprisoned the brothers Nephi and Lehi. The Lamanites cried out, "What shall we do, that this cloud of darkness may be removed from overshadowing us?" The inspired answer instructed: "You must repent, and cry unto the voice, even until ye shall have faith in Christ, . . . and when ye shall do this, the cloud of darkness shall be removed from overshadowing you." God showed them that this cloud of physical darkness was like their spiritual darkness, but that they could remove it if they wished to (Helaman 5:40–43). When they followed instructions, the cloud was removed and a holy fire encircled every soul.

A second example appears in King Mosiah's observation, following the miraculous escape of King Limhi's people by the Lord's intervention: "But behold, he did deliver them because they did *humble* themselves before him; and because they *cried mightily* unto him he did deliver them out of bondage; and thus doth the Lord work with his power in *all cases* among the children of men, extending the arm of mercy towards them that *put their trust* in him" (Mosiah 29:19–20). The efficacy and power of humility and crying mightily to the Lord are taught repeatedly as the means by which one gains access to divine deliverance "in all cases."

One important means by which the Book of Mormon makes divine deliverance understandable to us is through accounts of journeys, such as the exodus of the children of Israel out of Egypt, a journey referred to by prophets throughout the Book of Mormon. These prophets used the Exodus as the prototype of deliverance, usually for the purpose of bringing the people to repentance through remembrance of God's miraculous deliverance in the past. Wherever the Exodus appears in the Book of Mormon, it appears within the larger context of deliverance.

For example, Nephi urged his brothers to help build the ship. He recounted the Exodus to enlist their confidence and cooperation (1 Nephi 17:23–31). He reminded them that on the Israelite journey, God fed his people manna (v. 28), that he caused water to come from the rock (v. 29), that he provided light and direction, and that he did "all things for them which were expedient for man to receive" (v. 30). Nephi compared the Exodus to the journey on which the Nephites were about to embark. He told his brothers that on this journey, too, God would be their light (v. 13), would make their food sweet (v. 12), and would provide every necessary thing for the journey—if they would keep his commandments. After all, the Lord was trying to bring them into alignment with those eternal principles that call forth prosperity. The Lord promised, Nephi explained further, that "after ye have arrived in the promised land, ye shall know that I, the Lord, am God; and that I, the Lord, did *deliver* you from destruction; yea, that I did bring you out of the land of Jerusalem" (v. 14). The point of allowing people to undertake journeys seems to be to make possible certain experiences that promote the discovery of the Lord's delivering power.

All major journeys in the Book of Mormon are allegorical as well as actual, and they reflect not only the different kinds of the Lord's deliverances but also the principles on which the deliverances depend. All these journeys typify every person's sojourn on earth and the tasks that each is given to accomplish. Only God has the overview of the journey, and only God knows what will be needed along the way. He offers everything each one needs to succeed in the quest. As the Book of Mormon amply illustrates, however, people must often be persuaded to receive Christ's divine deliverance for their earthly journeys.

The destination of each divinely guided journey is a promised land where spiritual enlargement will be possible. The land prepared by God is "a land which is choice above all other lands" (1 Nephi 2:20; see also Ether 1:42). And, as the journeys represent the individual's sojourn on earth, so the destinations represent the kingdom of heaven, or reentering the presence of God. Again, the journeys represented in the Book of Mormon typify everyone's earthly sojourn and his or her need for divine help at every juncture.

Four examples will suffice to illustrate the principles underlying deliverance on journeys:

1. Lehi's journey to the New World.
2. Alma the Elder's journey from the land of Nephi across the wilderness to the land of Zarahemla.
3. The trek of King Limhi and Ammon to Zarahemla.
4. The Jaredite voyage to the choice land.

The first example is Lehi's journey. On their way to a promised land, Lehi and his family began a seemingly impossible trip through dangerous wilderness and across a terrifying ocean. Alma provided the allegorical interpretation of this journey and emphasized both the necessity as well as the ease of consulting the Lord in all our affairs (Alma 37:38–47). Here in Alma we learn the name of Lehi's ball or director, *Liahona,* which signifies a compass (v. 38). Because the availability of revelation is a difficult spiritual reality for people to grasp, the Lord designed the palpable Liahona not only to help Lehi's family find their way to the promised land but also to teach the principles which govern revelation, illustrating how individuals actually go forward depending on God as though they were holding a Liahona in their hands. Nephi explained how they made the compass work:

"And it came to pass that I, Nephi, beheld the pointers which were in the ball, that they did work according to the faith and diligence and heed which we did give unto them.

"And there was also written upon them a new writing, which was plain to be read, which did give us understanding concerning the ways of the Lord; and it was written and changed from time to time, according to the faith and diligence which we gave unto it. And thus we see that by small means the Lord can bring about great things" (1 Nephi 16:28–29).

Alma explained further:

"And it did work for them according to their faith in God; therefore, if they had faith to believe that God could cause that those spindles should point the way they should go, behold, it was done; therefore they had this miracle, and also many other miracles wrought by the power of God, day by day.

"Nevertheless, because those miracles were worked by small means it did show unto them marvelous works. They were slothful, and forgot to exercise their faith and diligence and then those marvelous works ceased, and they did not progress in their journey;

"Therefore, they tarried in the wilderness, or did not travel a direct course, and were afflicted with hunger and thirst, because of their transgressions" (Alma 37:40–42).

We learn at least four simple but profound principles here that teach us how to go to the Lord for help:

If they just *believed* that the ball would deliver them, it did. Simple belief connects the believer with the powers of heaven.

Their belief not only made the ball work but also made it possible for them to receive many other miracles, even day by day. Once one begins to exercise belief, he creates a new dimension for himself in which frequent miracles can occur.

Although the means were small, the works were marvelous. As a ship is worked by a small helm (D&C 123:16), so the powers of divine deliverance are engaged by small means on earth: that is, by belief, humility, humble petitioning of the Lord, obedience, and persistence.

When the travelers grew lazy and neglected to ask, divine deliverance ceased, and they became hungry, thirsty, and lost. The Lord requires focused energy of us—faith, diligence, giving heed. So simple, but so powerful.

Alma explained the symbolism of the Liahona's delivering power: "I would that ye should understand that these things are not without a shadow; for as our fathers were slothful to give heed to this compass (now these things were temporal) they did not prosper; even so it is with things which are spiritual.

"For behold, it is as easy to give heed to the word of Christ, which will point to you a straight course to eternal bliss, as it was for our fathers to give heed to this compass, which would point unto them a straight course to the promised land.

"And now I say, is there not a type in this thing? For just as surely as this director did bring our fathers, by following its course, to the promised land, shall the words of Christ, if we follow their course, carry us beyond this vale of sorrow into a far better land of promise" (Alma 37:43–45).

We see here that the Lord provides physical, tangible symbols to teach spiritual realities and to help the reader understand unseen spiritual powers. The tools of deliverance are interesting in themselves: the Liahona, the plates of brass, the sword of Laban, fire, clouds, boats, and shining stones. Each instrument of deliverance represents the unseen but real, accessible power of miracles in the Savior.

The second and third examples of journeys, both recorded in Mosiah, illustrate again the conditions on which divine deliverance is granted. God provides deliverance in response to the preparation and righteousness of the people. For example, in the case of Alma's group in Helam, the people escaped during broad daylight as the enemy miraculously slept, in contrast to the more natural escape of Limhi's community, which took place under cover of night while drunken Lamanites slept (Mosiah 23–24). Clearly some deliverances happen miraculously, whereas others occur more naturally and progress more slowly; still, they come by the Lord's power. In the case of Limhi's group, the people needed more time to repent of Abinadi's martyrdom before they were ready for deliverance, and so the Lord took more time to respond: "The Lord was slow to hear their cry because of their iniquities; nevertheless the Lord did hear their cries, and *began* to soften the hearts of the Lamanites that they *began* to ease their burdens; yet the Lord did not see fit to deliver them out of bondage" (Mosiah 21:15).

Alma's group, on the other hand, had believed on Alma's words alone, and had left their property and risked their lives to be baptized; therefore, they were prepared to exercise more faith and to accept a more miraculous deliverance: "Alma and his people did not raise their voices to the Lord their God, but did pour out their hearts to him; and he did know the thoughts of their hearts. And it came to pass that the voice of the Lord came to them in their afflictions, saying: Lift up your heads and be of good comfort, for I know of the covenant which ye have made unto me; and I will covenant with my people and deliver them out of bondage . . . that ye may know of a surety that I, the Lord God, do visit my people in their afflictions. . . . [A]nd they did submit cheerfully and with patience to all the will of the Lord. And it came to pass that so great was their faith and their patience that the voice of the Lord came unto them again, saying: Be of good comfort, for on the morrow I will deliver you out of bondage" (Mosiah 24:12–16). The Lord suits the type of deliverance to the spiritual needs of the groups involved.

The fourth example is the Jaredite journey across the ocean. This journey provides another example of physical and spiritual deliverance. Tangible instruments of deliverance abound here. Moroni recorded that when the Jaredites crossed the great deep in their watertight vessels, the Lord "caused that there should be a furious wind blow upon the face of the waters, towards the promised land. . . . [M]any times [they were] buried in the depths of the sea, because of the mountain waves which broke upon them, and also the great and terrible tempests which were caused by the fierceness of the wind. . . . [W]hen they were encompassed about by many waters they did cry unto the Lord, and he did bring them forth again upon the top of the waters," and "they were driven forth; and no monster of the sea could break them, . . . and they did have light continually, whether it was above the water or under the water" (Ether 6:5–7, 10).

The recurrent motif of light in these journeys, and in this case from shining stones, draws our attention. During the Exodus, Jehovah led the children of Israel by a pillar of fire. The Lord had earlier told the brother of Jared, "For behold, I am the Father, I am the *light,* and the life, and the truth of the world" (Ether 4:12). In the course of the terrifying journey, these Jaredites could see the light from the stones and understand that it represented the unseen love of Jesus Christ. The journey through the deep also recalls the Savior's teaching about the winds and rains that beat vainly upon the invincible man or woman of Christ (Matthew 7:24–25).

The Book of Mormon offers help from personal trouble. Nephi, angry and in despair, cried out: "O wretched man that I am! Yea, my heart sorroweth because of my flesh; my soul grieveth because of mine iniquities. I am encompassed about, because of the temptations and the sins which do easily beset me. And when I desire to rejoice, my heart groaneth because of my sins" (2 Nephi 4:17–19).

But as his heart turned to many evidences in his own life of the Lord's love and intervention, he rebuked himself for his despair, because he remembered the principle of deliverance. Nephi's is perhaps the most sublime expression in scripture of faith in the Savior's power to deliver:

"Awake, my soul! No longer droop in sin. Rejoice, O my heart, and give place no more for the enemy of my soul.

"Do not anger again because of mine enemies. Do not slacken my strength because of mine afflictions.

"Rejoice, O my heart, and cry unto the Lord, and say: O Lord, I will praise thee forever; yea, my soul will rejoice in thee, my God, and the rock of my salvation.

"O Lord, wilt thou redeem my soul? Wilt thou *deliver* me out of the hands of mine enemies? Wilt thou make me that I may shake at the appearance of sin? . . .

"Yea, I know that God will give liberally to him that asketh. Yea, my God will give me, if I ask not amiss; therefore I will lift up my voice unto thee; yea, I will cry unto thee, my God, the rock of my righteousness. Behold, my voice shall forever ascend up unto thee, my rock and mine everlasting God" (2 Nephi 4:28–31, 35).

Moroni taught that despair comes of iniquity (Moroni 10:22). By *iniquity* he seems to mean lack of faith in the deliverance offered by the Savior. He stated, "Christ truly said . . . : If ye have faith ye can do all things which are expedient unto me" (Moroni 10:23). That is, because there is a Savior, there are solutions to seemingly insolvable problems.

The life of Alma the Younger demonstrates several examples of individual deliverance. He declared that he was "supported under trials and troubles of every kind, yea, and in all manner of afflictions; . . . and I do put my trust in him, and he will still *deliver* me" (Alma 36:27). Alma gave the benefit of his belief and experience to his son: "I would that ye should remember, that as much as ye shall put your trust in God even so much ye shall be *delivered* out of your trials, and your troubles, and your afflictions, and ye shall be lifted up at the last day" (Alma 38:5). Although in the following passage he did not use the word *deliverance,* he clearly described a release from his own personal hell: "For three days and for three nights was I racked, even with the pains of a damned soul.

" . . . I was thus racked with torment. . . .

" . . . I cried within my heart: O Jesus, thou Son of God, have mercy on me, who am in the gall of bitterness, and am encircled about by the everlasting chains of death.

"And now, behold, when I thought this, I could remember my pains no more; yea, I was harrowed up by the memory of my sins no more.

"And oh, what joy, and what marvelous light I did behold; yea, my soul was filled with joy as exceeding as was my pain!" (Alma 36:16–20).

Later, as a mature missionary, Alma viewed the abysmal apostasy of the Zoramites and exclaimed: "O Lord, my heart is exceedingly sorrowful; wilt thou comfort my soul in Christ. . . . O Lord, wilt thou comfort my soul, and give unto me success." Then, speaking for his companions, he said: "Yea, wilt thou comfort their souls in Christ" (Alma 31:31–32). "And the Lord provided for them that they should hunger not, neither should they thirst; yea, and he also gave them strength, that they should suffer no manner of afflictions, save it were swallowed up in the joy of Christ" (Alma 31:38). Thus Alma impresses us with the point that divine deliverance is readily available to those who will come to the Lord.

All the dilemmas illustrated in the Book of Mormon contain dangerous elements uncontrollable by mortals, so that when deliverance comes, no one will be confused about the One from whom it comes. Life's path is strewn with seemingly unsolvable dilemmas so that people will be driven to God for help. The Lord's methods may be based on the principle that the greater the trouble, the more likely one will turn to him for help. We are reminded that the only way that God can teach how faith works is through experience, some of it necessarily dangerous. When the hand of God is revealed in the midst of a seemingly unsolvable situation, one's faith can become unshakable.

If the Book of Mormon is really about deliverance, it is also about Christ's atonement. Therefore, every instance of *deliverance* is also an instance of redemption, salvation, and at-one-ment. The Book of Mormon was provided, at least in part, to illustrate how grace and atonement actually work in the lives of those who come to Christ.

Obviously the trek of Limhi and his people across a wilderness some two thousand years ago may mean little to the reader—until he realizes that Limhi's journey is analogous to his own life journey. Thereafter, a person will read the Book of Mormon differently as he grasps the insight that humility, prayer, and obedience can draw down divine deliverance in the midst of one's own wilderness trials. The Book of Mormon is a handbook of principles for traveling one's earthly path by the divine enabling power of the Lord Jesus Christ. The Book of Mormon is itself a tool of deliverance. Nephi made the same point with this instruction: "Wherefore, I said unto you, feast upon the words of Christ; for behold, the words of Christ will *tell you all things what ye should do*" (2 Nephi 32:3).

We have seen that the instances of deliverance throughout the Book of Mormon can infuse us with hope for deliverance from our own troubles, instruct us in how to come to the Lord for help, and fill the soul with faith in the eternal constancy and accessibility of the great Deliverer.

Notes

From *Doctrines of the Book of Mormon: The 1991 Sperry Symposium*, ed. Bruce A. Van Orden and Brent L. Top (Salt Lake City: Deseret Book Co., 1992), 182–93.

1. This figure also includes a few instances of *deliver* to mean "to hand over" as in "the Lord will deliver Laban into your hands" (1 Nephi 3:29) or "I did deliver the plates unto my brother Chemish" (Omni 1:8). Obviously synonyms like *save* and *preserve* might be studied in combination with *deliverance*. My objective here is not a word study but a demonstration of how the Lord used one word to make clearer the abstract principle of God's grace.

2. The emphasis in this verse and in subsequent scriptural passages in this chapter has been added by the author.

3. For discussion of Exodus language used in the Book of Mormon, see S. Kent Brown, "The Exodus Pattern in the Book of Mormon," *BYU Studies* 30, no. 3 (Summer 1990): 111–26.

JACOB'S ALLEGORY:
THE MYSTERY OF CHRIST

Joseph Smith explained the way to understand parables and allegories: "I have a key by which I understand the scriptures. I enquire, what was the question which drew out the answer?"[1] Jacob poses two key questions in the introduction to the allegory of the olive tree (Jacob 5) that provide some clues to its meaning. First, Jacob asks: "Why not speak of the atonement of Christ, and attain to a perfect knowledge of him?" (Jacob 4:12).[2] Jacob then points to the Jews' deliberate efforts to distance God and render him incomprehensible: they sought to create a god who could not be understood (Jacob 4:14). For their self-inflicted blindness God took away "his plainness from them . . . because they desired it" (Jacob 4:14). Here Jacob asks the second key question: "My beloved, how is it possible that these [the Jews], after having rejected the sure foundation, can ever build upon it, that it may become the head of their corner? Behold, my beloved brethren, I will unfold this mystery unto you" (Jacob 4:17–18). Among other meanings, a mystery is a spiritual truth grasped only through divine revelation. The mystery that Jacob unfolds, therefore, counters the Jews' deliberate mystification of God and reveals the true nature of Jesus Christ and his divine activity in the lives of even the most intractable of men. Jacob's two key questions alert the reader that the allegory will deal with grace, Atonement, and the relationship of these principles to Israel.

Superficially, the allegory is the story of a man and his olive tree and the man's efforts to restore the deteriorating tree to its former pristine condition. At a deeper level, the allegory treats God's response to Israel's spiritual death, represented by the geographically scattered condition of the tree's branches. The state of separation of the people of Israel from each other indicates that the Atonement is not working in their lives; otherwise, they would live in Zion together. The allegory describes God's efforts to gather these disparate parts of Israel into at-one-ment with him. Fifteen times we read that he wishes to preserve the harvestable fruit and lay it up, as he says, "to mine own self."

In Latter-day Saint usage, *atonement,* or *at-one-ment,* refers not only to the act of redemption Jesus wrought in Gethsemane and on the cross, but also to the Lord's ongoing labors to bring his children back into oneness with him. After all, it is his work, as well as his glory, to bring to pass the eternal life of man (Moses 1:39). The word *atonement* first appears in William Tyndale's 1526 English version of the Bible.[3] He used the word *at-one-ment* to translate the Greek word for *reconciliation* (*katalage*) as found in Romans 5:11. The Savior's yearnings for this state of oneness with his children appear not only in this allegory but also in such places as the great intercessory prayer in John 17 and the luminous prayer sequences in 3 Nephi 19. In seeking an understanding of Jacob's allegory, it is helpful to understand the strength of the divine desire behind the process of at-one-ment.

We approach the meaning of the Atonement in Jacob's allegory by consulting scripture for additional references to trees. Scripture abounds with symbolic trees. A tree planted by a river is an Old Testament symbol of a righteous man (Psalm 1:3; Jeremiah 17:8). Isaiah writes, "The Lord hath anointed me to preach good tidings unto the meek; . . . that they might be called trees of righteousness" (Isaiah 61:1, 3). In Daniel's dream a great tree represents a man (Daniel 4:10, 22). Another tree in Isaiah produces a stem (of Jesse), which is Christ (D&C 113:1–2). Two famous trees grow in Eden: the tree of knowledge of good and evil and the tree of life (Genesis 2:9,

17). We might gather from Alma 42:6–7 that the tree of life represents the presence of God. A millennial tree of life in Revelation 22:2 has leaves to heal the nations, an obvious reference to the Savior. Jesus is hanged on a tree of life (Acts 5:30). In vision Lehi and Nephi see a divine tree that pertains to Jesus' ministry of love (1 Nephi 8:10; 11:8). Lehi's dream tree receives at least three meanings: the Son of God and his divine activity (1 Nephi 11:7); the love of God (1 Nephi 11:22, 25); and the tree of life (1 Nephi 11:25; 15:22). Since these meanings all overlap, we would understand that Lehi's dream tree represents multiple facets of Christ.

Most often in scripture, then, the tree is an anthropomorphic symbol. A tree serves well as such a symbol because it has, after all, limbs, a circulatory system, the ability to bear fruit, and so forth. Specifically, scriptural trees stand either for Christ and his attributes or for man.

Here we might make an observation about divine symbols. The finite mind wants to pin down a one-to-one correspondence between the elements of an allegory and that which the elements represent. But the divine mind works in multiple layers of meanings for symbols. In scripture the meaning often lies in the *aggregate* of allusions and associations. The olive tree is one of these layered symbols. It is Israel at the macrocosmic level; it is also an individual Israelite being nourished by an attentive God.

But the olive tree seems also to reflect the Savior himself, as we can see when we analyze the relationship between Jacob's olive tree and Lehi's dream tree. The two trees appear in juxtaposition with each other in 1 Nephi chapters 8 through 15. Lehi's dream tree first appears in chapter 8. The first reference to the olive tree appears two chapters later, in chapter 10, the grafting in to this olive tree being defined as coming to the knowledge of the true Messiah (1 Nephi 10:12–14). Then in chapter 11 Lehi's dream tree is shown to Nephi, who observes that the tree is the Son of God shedding forth his love (1 Nephi 11:7, 21–22). Next, in chapter 15, Nephi explicates the olive tree for his brethren, saying that the covenant people will receive strength and nourishment from the *true* vine when they are grafted into the *true* olive tree (1 Nephi 15:16). The reference to the true vine suggests a passage from John: "I [Christ] am the vine, ye are the branches: He that abideth in me, and I in him, the same bringeth forth much fruit: for without me ye can do nothing" (John 15:5). This discussion of the true vine and the *true* olive tree leads to Nephi's explication of the dream tree, suggesting that a strong relationship between these two trees exists in the minds of both Lehi and Nephi, since they are discussed alternately. Thus the dream tree is Christ, and the true olive tree is Christ.

Extending this point, we can examine the fruit of these two trees. When Jacob is about to introduce the allegory he exhorts the reader to be the *first-fruits* of Christ (Jacob 4:11). Nephi says that the fruit from Lehi's dream tree is "*most precious* and most desirable above all other fruits" (1 Nephi 15:36). In identical language, the olive tree's natural fruit is "*most precious* above all other fruit" (Jacob 5:61) and "*most precious* unto him from the beginning" (Jacob 5:74); that is, the fruit from both trees is described as "most precious." It would seem that the fruit represents harvestable souls, or those that can be or have been sanctified by the Savior's atoning power. Both the olive tree and the dream tree in their sanctified state are the same tree, and the merging of these trees through these chapters heightens the message of at-one-ment between man and Christ. At the end of time, all of the trees and fruits have merged. The Lord observes that the trees have become "like unto one body; and the fruits . . . equal" (Jacob 5:74).

While we consider the olive tree, we may also wish to examine olive oil for its relationship to Atonement. In ancient Israel, the olive tree was the tree of life; olive oil was used in sacrifices and in ritual purification, rites that symbolized the restoration of God's favor and the return of joy to a man previously disgraced.[4] It was associated with vigor and fertility. The sick were anointed with oil. Brides were anointed prior to marriage. Anointing with oil and washing and dressing symbolized a change of status throughout the Old Testament; for example, the consecration of Aaron to the priesthood included washing, donning of special garments, and anointing his head with oil (Leviticus 8:6–12). The holy anointing oil, which could not be used for any profane purpose, was prepared by Moses in the desert and was kept in the Holy of Holies, serving for the anointing of the Tabernacle and of all high priests and kings (Exodus 30:25–31). Prophets were anointed with oil, as were temples and altars (Genesis 28:18). Olive oil was indispensable in the preparation of the Passover lamb.[5] We remember that Christ is the *Anointed One*.

Perhaps the ultimate definition of oil in scripture, that which draws together all those mentioned above, appears in the Savior's parable of the Ten Virgins (Matthew 25:1–13), which he explicates in Doctrine and Covenants 45, identifying the oil as the Holy Ghost (D&C 45:56–57). The Lord Jesus is the agent of the Atonement, but the medium of the at-one-ment is the Holy Ghost—that sap or moisture that flows from the trunk through the branches. Perhaps something of this idea suggests itself in Jacob 5:18: "The branches of the wild tree have taken hold of the *moisture* of the root thereof, that the root thereof hath brought forth much . . . tame fruit." Jacob makes a similar metaphorical connection when he exhorts Israel not to "quench the Holy Spirit" (Jacob 6:8).

Though the symbolic elements of the allegory represent historical people and events, a yet greater insight may lie in the allegory as a theodicy, that is, God's explication of himself, his work, and his love. Not only did the Jews dematerialize God and scramble the facts about him, but so also has nearly every apostasy since. The mystery that Jacob illuminates is that God is not distant, but full of grace— of divine enabling power—ceaselessly involving himself with each of his children, seeking a response, seeking a relationship.

It is in the figures of pruning, grafting, and digging about that the Lord reveals most specifically the function of the at-one-ment. The allegory describes this divine activity as wrought both in the tree and in the environment of the tree, suggesting that God seeks access to man at several points. *Grafting in* might represent events and experiences that bring one to Christ—conversion. *Digging about* suggests the divine structuring of one's environment for individual tutorials. *Dunging* suggests spiritual nourishing. As to *pruning,* we might understand those painful experiences in which we feel stymied as we pit our self-will against the Lord (Mosiah 7:29). Hugh B. Brown provided an excellent illustration in his little parable of the currant bush. At the end the Gardener speaks to the little bush, which he has cut back again and again:

"Do not cry . . . what I have done to you was necessary . . . you were not intended for what you sought to be, . . . if I had allowed you to continue . . . you would have failed in the purpose for which I planted you and my plans for you would have been defeated. You must not weep; some day when you are richly laden with experience you will say, 'He was a wise gardener. He knew the purpose of my earth life. . . . I thank him now for what I thought was cruel.'" The current bush then cries, "Help me, dear God, to endure the pruning, and to grow as you would have me grow; to take my allotted place in life and ever more to say, 'Thy will not mine be done.'"[6]

This ceaseless divine activity in seeking to bring men into the Lord's presence, even while they walk the earth, is reflected in the continual nourishing, digging, and pruning going on in the allegorical vineyard. The word *nourish* appears twenty-one times in the seventy-seven verses of the chapter, along with the words *digging, dunging, pruning,* and *preserving,* which appear frequently along with *nourishing,* indicating that the idea of nourishing, of personal attention to Israel and to the Israelites, is a major theme of the allegory. The perfect knowledge of Christ that Jacob refers to (Jacob 4:14), that is, at-one-ment with him, is achieved in Christ's revelation of himself through the pruning, digging, and nourishing of his individual covenant children.

The idea that God himself seeks continual association with each of his covenant children is expressed in other Book of Mormon passages. Alma declared, "A shepherd hath called after you and is *still* calling after you. . . . The good shepherd doth call you; yea, and in his own name he doth call you, which is the name of Christ" (Alma 5:37–38). Lehi exclaimed, "I am encircled about eternally in the arms of his love" (2 Nephi 1:15). Helaman wrote to Moroni: "May the Lord our God, who has redeemed us and made us free, keep you continually in his presence" (Alma 58:41). Christ spoke poignantly in Revelation, "Behold I stand at the door, and knock" (Revelation 3:20).

If God is seeking access to his children continually, what is the meaning of the periods of divine absence in the allegory? The Lord declares, "I have stretched forth mine hand *almost* all the day long" (Jacob 5:47). Jacob drops the word *almost* when he reiterates: "He stretches forth his hands unto them *all* the day long. . . . Come with full purpose of heart, and cleave unto God as he cleaveth unto you. . . . For why will ye die? . . . For behold, . . . ye have been nourished by the good word of God *all* the day long" (Jacob 6:4–7). *Cleave* is Atonement language. It is not God who has ceased to cleave, but man who has rejected God's love. These periods in which we do not see divine activity signify not so much the Master's absence, but rather *Israel's* voluntary withdrawal from the true olive tree.

At the end of the allegory the Lord speaks to his servant, "Blessed art thou; . . . because ye have been diligent in laboring with me in my vineyard, . . . ye shall have joy with me because of the fruit of my vineyard" (Jacob 5:75). Jacob echoes these words: "How blessed are they who have labored diligently in his vineyard" (Jacob 6:3)—those who have participated in the divine activity of at-one-ment. In latter days the Lord has said, "I will gather together in *one* all things, both which are in heaven, and which are on earth" (D&C 27:13). The allegory underscores the fact that the greatest work going forth on the earth is the work of bringing those who are scattered, alienated, and miserable back into harmony and oneness with each other and with the Creator.

One of the key insights that emerges from the allegory is that the power of the Atonement seeks to affect men at every level of their existence. It urges people together geographically into Zions. It promotes generosity and consecration of goods. It prompts people to resonate emotionally and to synergize spiritually. The Lord says, "I say unto you, be one; and if ye are not one ye are not mine" (D&C 38:27).

Finally, an individual must discover Jacob's mystery for himself.

The greatest value of the allegory may be that it serves to make one conscious of the efforts of the Lord to draw him by "the enticings of the Holy Spirit" (Mosiah 3:19) into a working relationship with a powerful Benefactor. This approach to the allegory enlarges one's confidence in the Lord's unceasing labors in his behalf and prompts him to search within to find the evidences of divine instruction and nurturing. The allegory teaches that the structure of oneness, of at-one-ment, is already in place. One need only discover and embrace the relationship.

Notes

Revised from "Jacob's Allegory: The Mystery of Christ," in *The Allegory of the Olive Tree,* ed. Stephen D. Ricks and John W. Welch (Salt Lake City: Deseret Book Co., 1994), 11–20.

1. Joseph Smith, *History of the Church of Jesus Christ of Latter-day Saints,* ed. B. H. Roberts, 7 vols. (Salt Lake City: Deseret Book Co., 1980), 5:261.
2. The emphasis in this verse and in subsequent scriptural passages in this chapter has been added by the author.
3. *Oxford English Dictionary* (Oxford: Oxford University Press, 1971), s.v. "atone," "atonement."
4. *Encyclopedia Judaica,* 17 vols. (Jerusalem: Keter, 1972), s.v. "oils," 12:1347.
5. *The Jewish Encyclopedia,* 12 vols. (New York: Funk & Wagnalls, 1901), s.v. "oil," 9:392.
6. Hugh B. Brown, *Eternal Quest* (Salt Lake City: Bookcraft, 1936), 245.

BENJAMIN AND THE MYSTERIES
OF GOD

K ing Benjamin had been praying for his people; in response, an
angel appeared with an important announcement. The king then
summoned his people to the temple to receive the angel's message in
connection with a sacred name. The people embraced the angel's
message, were born again, and entered into a holy covenant.

In this simple statement of basic facts from the Book of Mormon
account, we discover at least four interesting questions: (1) What was
Benjamin's role in the rebirth experience at the temple, and for what
was he praying? (2) What was it the angel came to announce? (3) How
would the name the king gave his people distinguish them from earlier
Nephites, who, for five hundred years, had anticipated the coming of
Christ? (4) What was the nature of the change that the people received,
and what does it all have to do with the mysteries of God? In the
pursuit of answers to these questions, we will explore the nature of
priesthood and its relationship to the mystery of spiritual rebirth.

THE MYSTERIES OF GOD

The scriptures repeatedly invite the reader to inquire about and
receive an understanding of the mysteries of God (Alma 26:22; D&C
6:11; 42:61). *Mysteries* are spiritual realities that can be known and
understood only by revelation, because they exist outside man's

sensory perception; but our scriptures record them, our prophets teach them, and the Holy Ghost reveals them to the diligent seeker. In fact, the whole gospel is a collection of mysteries—truths pertaining to salvation that would not be known by men in the mortal probation if God did not reveal them. Benjamin's address begins with an invitation to prepare to view the mysteries of God:

"My brethren, all ye that have assembled yourselves together, . . . I have not commanded you to come up hither to trifle with the words which I shall speak, but that you should hearken unto me, and open your ears that ye may hear, and your hearts that ye may understand, and your minds that the *mysteries of God* may be unfolded to your view" (Mosiah 2:9).[1]

The particular mystery that draws our attention here is the mystery of spiritual rebirth and the role that Benjamin's priesthood office played in that experience. With respect to the revelation of mysteries and the power of the priesthood, the Lord has said: "The power and authority of the higher, or Melchizedek Priesthood, is to hold the keys of all the spiritual blessings of the church—to have the privilege of receiving the mysteries of the kingdom of heaven, to have the heavens opened unto them, to commune with the general assembly and church of the Firstborn, and to enjoy the communion and presence of God the Father, and Jesus the mediator of the new covenant" (D&C 107:18–19).

In the first part of this chapter we will examine Benjamin's priesthood role as a prelude to understanding his prayer, the angel's response, and the spiritual rebirth of his people.

BENJAMIN'S PRIESTHOOD ROLE: THE POWER TO BLESS

A study of Benjamin's role gives opportunity to look at Benjamin's priesthood work in particular, but also at priesthood in general. Priesthood is the great governing authority of the universe. It unlocks spiritual blessings of the eternal world for the heirs of salvation. The power to play a saving role is the most sought-after power among righteous priesthood holders in time or in eternity. "What was the power of Melchisedek?" Joseph Smith asked.

"'Twas not the priesthood of Aaron etc., [but it was the power of] a king and a priest to the most high God. [That priesthood was] a perfect law of Theocracy holding *keys of power and blessings*. [It] stood as God to give laws to the people, administering endless lives to the sons and daughters of Adam."[2]

The Prophet Joseph further explained the centrality of priesthood in the eternal plan: "[Priesthood] is the channel through which all knowledge, doctrine, the plan of salvation and every important matter is revealed from heaven. . . . It is the channel through which the Almighty commenced revealing his glory at the beginning of the creation of this earth and through which he has continued to reveal himself to the children of men to the present time and through which he will make known his purposes to the end of time."[3]

A brief look at the history of the priesthood on the earth reveals that men like Benjamin have stood in this priesthood channel unlocking the blessings of salvation for their people since the days of Adam. Adam, in fact, was the great prototype of priesthood holders who strove to bring their communities and their posterity into at-one-ment with the Lord Jesus Christ. Adam blessed his posterity because, the Prophet Joseph taught, "he wanted to bring them into the presence of God. They looked for a city . . . 'whose builder and maker is God.' (Hebrews 11:10.)"[4]

After Adam, Enoch labored with his people and succeeded in bringing to pass not only their sanctification but also their translation (Moses 7:21), a function of the higher priesthood. Following Enoch, Melchizedek, king and high priest of Salem, brought many into the fulness of the priesthood and the presence of God. His people also received translation, "obtained heaven, and sought for the city of Enoch" (JST, Genesis 14:34).

After Melchizedek, Moses strove for the same blessings for his people. The Lord said to them: "If ye will obey my voice indeed, and keep my covenant, then ye shall be a peculiar treasure unto me above all people. . . . And ye shall be unto me a kingdom of priests, and an holy nation" (Exodus 19:5–6), touching again on this idea of the holy city. Moses "sought diligently to sanctify his people that they might behold the face of God; but they hardened their hearts and could not endure his presence; therefore, the Lord . . . swore that they should not enter into his rest while in the wilderness, which rest is the fulness of his glory" (D&C 84:23–24). The Prophet Joseph explained: "Moses sought to bring the children of Israel into the presence of God, through the power of the Priesthood, but he could not."[5]

In this same tradition, Joseph Smith showed great anxiety to see the kingdom fully established and the temple completed before his death, saying: "Hurry up the work, brethren. . . . Let us finish the temple; the Lord has a great endowment in store for you, and I am anxious that the brethren should have their endowments and receive the fullness of the Priesthood. . . . Then . . . the Kingdom will be established, and I do not care what shall become of me."[6]

Our modern prophets strive in the same manner as Benjamin and all ancient prophets to sanctify the members of the Church and to unlock these priesthood powers in their behalf. Elder David B. Haight made reference to this power as he recounted a sacred experience in which he viewed the Savior's ministry and came to a greater understanding of the power of the priesthood:

"During those days of unconsciousness I was given, by the gift and power of the Holy Ghost, a more perfect knowledge of His mission. I was also given a more complete understanding of what it means to exercise, in His name, the *authority to unlock the mysteries of the kingdom of heaven* for the salvation of all who are faithful."[7]

Thus Benjamin, as a prophet, seer, revelator, king, and priest, held the keys of power and blessing for his community. A priesthood holder's office is to sanctify himself and stand as an advocate before God seeking blessings for his community in the manner of the Lord Jesus Christ himself (John 17:19), whether the community be as small as a family or as large as a kingdom. A righteous priesthood holder can work by faith to provide great benefits to his fellow beings (Mosiah 8:18). He can, in fact, exercise great faith in behalf of others of lesser faith, "filling in" with faith for them; thus a prophet and a people together can bring down blessings for even a whole community (for example, see Ether 12:14). The Lord seems to be interested not only in individuals but also in groups of people who wish to establish holy cities and unite with heavenly communities. Like the ancients, one who holds the holy priesthood is always trying to establish a holy community, is always "look[ing] for a city" (Hebrews 11:10, 16). So it was with Benjamin.

PRIESTHOOD POWER OVER ENEMIES

In analyzing the scriptural accounts of priesthood work, we discover that one major task of priesthood holders, in unlocking the blessings of salvation for their people, is to triumph over the powers of evil—over "enemies." Of this task, Joseph Smith said: "Salvation is nothing more nor less than to triumph over all our enemies and put them under our feet. And when we have power to put all enemies under our feet in this world, and a knowledge to triumph over all evil spirits in the world to come, then we are saved."[8]

This, then, is the pattern: the priesthood holder labors with all his faculties to rout Satan from his loved ones as that enemy is manifested in contention, mental warfare, and physical violence among the people. For any Melchizedek Priesthood holder to become a prince of peace, he must in some degree wrest his kingdom, great or small, from the adversary and halt the plans of the destroyer on behalf of his loved ones. The Book of Mormon's description of Melchizedek reflects this pattern:

"Now this Melchizedek was a king over the land of Salem; and his people had waxed strong in iniquity and abomination; yea, they had all gone astray; they were full of all manner of wickedness; but Melchizedek having exercised mighty faith, and received the office of the high priesthood according to the holy order of God, did preach repentance unto his people. And behold, they did repent; and

Melchizedek did establish peace in the land in his days; therefore he was called the prince of peace" (Alma 13:17–18).

In this priesthood pattern, Benjamin labored against manifest evil and spiritual entropy to save his people. He waged war against the destroying Lamanites—using the sword of Laban, going forth in the strength of the Lord (Words of Mormon 1:13–14); he waged other battles against false prophets, preachers, and teachers; and then he put down contention among his people with the assistance of other holy prophets. The record says that "king Benjamin, by laboring with all the might of his body and the faculty of his whole soul, . . . did once more establish peace in the land" (Words of Mormon 1:18). Peace, that essential condition for spiritual progress, is evidence of the triumph of spiritual principle and also of the preparation of the people in any size group to receive greater spiritual blessings.

Benjamin, then, was not an anomaly; he acted in the tradition of all true prophets before and after him in unlocking spiritual blessings for his people as he strove to prepare them to return to God. He was therefore the very person in Zarahemla who had the power to pray that spiritual blessings be poured out on this community of Saints that they might be born again.

BENJAMIN'S PEOPLE, THE ANGEL, AND THE SPIRITUAL REBIRTH

Benjamin's people were not spiritual novices. They were descendants of those righteous Nephites who fled from the land of Nephi under Benjamin's father, King Mosiah$_1$, and were led by the power of God to the land of Zarahemla, where they united with Zarahemla's people. The first Mosiah, whom I will designate as Mosiah$_1$ restored the gospel among them and reigned over them. The important point here is that Benjamin's people were not spiritually ignorant; they were not hearing about the Lord Jesus Christ on this occasion for the first time. The record states clearly that they were "a diligent people in keeping the commandments of the Lord" (Mosiah 1:11); it states that there were not any among them, except little children, who had not been taught "concerning the . . . prophecies which have been spoken by the holy prophets" and all that the Lord commanded their fathers to speak (Mosiah 2:34; see also v. 35). I assume here that Benjamin's people, having been taught the gospel, had been previously baptized in the name of Jesus Christ.

In addition, we might infer that Benjamin's people came up to the temple with some preparation for and in some anticipation of a spiritual event. They would have been aware of what their king had been trying to do for them according to the ancient pattern. They knew there was a blessing awaiting them. They came up to the temple, in part, to give thanks to God for their king "who had taught them to keep the commandments of God, that they might rejoice and be filled with love towards God and all men" (Mosiah 2:4).

The phrases *to rejoice and be filled with love* and *to be filled with joy* seem to have a technical meaning in scripture. They appear to be alternative ways of describing being *born again*. Scripture abounds with references to being filled with this transforming joy and love under the influence of the Holy Ghost. Nephi said, for example, "[God] hath *filled me with his love,* even unto the consuming of my flesh" (2 Nephi 4:21); the Lamanites were "*filled with that joy* which is unspeakable and *full of glory.* . . . The Holy Spirit of God did come down from heaven, and did enter into their hearts, and they were *filled as if with fire*" (Helaman 5:44–45); the Nephites with the resurrected Christ "were *filled with the Holy Ghost* and with fire" (3 Nephi 19:13); Mormon taught us to pray to be *filled with this love,* which is charity, or perfect love, which makes one pure like Christ (Moroni 7:48).

Compare now the experience of Benjamin's people: "The Spirit of the Lord came upon them, and they were *filled with joy,* having received a remission of their sins, and having peace of conscience, because of the exceeding faith which they had in Jesus Christ who should come, according to the words which king Benjamin had spoken unto them" (Mosiah 4:3).

It seems that being filled with joy, love, and glory are all ways of describing being born again. Benjamin clearly identified for the people what they had experienced: "Behold, this day he hath spiritually begotten you; for ye say that your hearts are changed through faith on his name; therefore, ye are born of him and have become his sons and his daughters" (Mosiah 5:7). He said that they had "come to the knowledge of the *glory of God*," as they "tasted of his love" (Mosiah 4:11).

One of the blessings of the priesthood is that it can bring others to be partakers of the glory of God. The Prophet Joseph taught that "being born again comes by the Spirit of God through ordinances."[9]

Angels, as priesthood holders, can also play a part. Indeed, as Alma taught, angels can be commissioned by God to cause "men to behold of [God's] glory" (Alma 12:29). Thus the angel said to Benjamin: "I am come to declare unto you the glad tidings of great joy. For the Lord *hath heard thy prayers,* and hath judged of thy righteousness, and hath sent me to declare unto thee that thou mayest rejoice; and that thou mayest declare unto thy people, that they may also be *filled with joy*" (Mosiah 3:3–4).

If to be *filled with joy* has close relationship to being *born again,* then it would seem that the angel had come from God to answer Benjamin's prayer that his people might be endowed with the name and receive the rebirth. The angel declared that the time had come that these people might literally be "filled with joy" (Mosiah 3:4) and that "whosoever should believe that Christ should come, the same might receive remission of their sins, and *rejoice with exceedingly great joy*" (Mosiah 3:13).

It was not just the news that the Savior would minister on the earth in the near future that filled them with joy—because they already knew all the prophecies of the holy prophets with respect to the Savior's ministry—but that the Atonement, the at-one-ment, was about to become very personal to *them.* Their faith in the Lord was about to become knowledge (Mosiah 5:4). This joy announced by the angel was not to be just a momentary experience or a memory that would fade with time. If they were diligent in prayer (Moroni 8:26) and obedient to other instructions their king would give them, they would be changed forever, could retain this perfect love and joy in their hearts, and would even "grow in the knowledge of the glory of [God]" (Mosiah 4:12). We might infer then that these two parties—the king and the people—had been praying and preparing for the time when the whole community, in the ancient tradition, might be redeemed and born again.

Without doubt, Benjamin knew what was going to transpire as he told his son to summon the people to the temple. He said, "I shall give this people a name . . . that never shall be blotted out, except it be through transgression" (Mosiah 1:11–12). Giving them the name forever is equivalent to causing them to be born again into the family of Christ. Because of the greater responsibility inherent in the formal taking of the name, Benjamin prefaced this spiritual endowment with warnings that if they proceeded with taking the name but then turned away in disobedience, they would have to drink of the cup of the wrath of God (Mosiah 3:26) and they would drink damnation to their souls (Mosiah 3:18, 25).

Notwithstanding the warning, the people crossed the threshold of spiritual experience into a fearsome, spiritually induced view of the reality of their fallen condition, confronting their own carnal state, even less than the dust of the earth. At the height of their distress, united under the influence of the Spirit, they cried aloud in the Lord's name and begged for a remission of sins (Mosiah 4:20). In response, the Spirit of the Lord descended upon them, and they were "filled with joy," the record says, fulfilling the exact words and promise of the angel (Mosiah 4:3). Their hearts were purified as they received a remission of their sins and peace of conscience because of their "exceeding faith . . . in Jesus Christ" (Mosiah 3:4). Benjamin observed, "He has poured out his Spirit upon you, and has caused that your hearts should be filled with joy, and has caused that your mouths should be stopped that ye could not find utterance, so exceedingly great was your joy" (Mosiah 4:20).

THE NATURE OF SPIRITUAL REBIRTH

In trying to comprehend the nature and extent of the spiritual experience described here in Mosiah, our own experience tells us, as do the scriptures, that spiritual experience can range from the subtle impressions of the Holy Spirit to dramatic encounters with heavenly powers. Thus, spiritual rebirth may begin at baptism, but without doubt additional degrees of spiritual rebirth and sanctification lie ahead for the true disciple, even a consummate change in which he has received the power to yield his heart entirely to God (Helaman 3:35). The description in Mosiah suggests such a change. In addition, based on other scriptures about being born again, it seems that the people partook to some degree of the following blessings:

As a result of the mighty change wrought in their hearts (Mosiah 5:2), they received Christ's image in their countenances; they could "sing the song of redeeming love," their hearts having been "stripped of pride" and enmity (Alma 5:26, 28; see also 5:12, 19).

Through the power of the Holy Ghost they were immersed in the heavenly fire, becoming one in God, attaining to a new order, as did Adam who, "born of the Spirit, and . . . quickened in the inner man, . . . heard a voice out of heaven, saying: Thou art baptized with fire, and with the Holy Ghost. . . . Thou art after the order of him who was without beginning of days or end of years, from all eternity to all eternity. Behold, *thou art one in me,* a son of God; and thus may all become my sons" (Moses 6:65–68).

To be born again is to be filled with "the Spirit of the Lord." Alma defined this mighty change when he proclaimed: "I had been born of God. Yea, and from that time even until now, I have labored without ceasing, that I might bring souls unto repentance; that I might bring them to taste of the exceeding joy of which I did taste; that they might also be born of God, and be *filled with the Holy Ghost*" (Alma 36:23–24).

They enjoyed a degree of sanctification (Mosiah 5:2). Their sins having been remitted, they could not look upon sin save with abhorrence; they also entered that spiritual dimension called "the rest of the Lord" (Alma 13:12).

The apostle John wrote, "Whosoever is born of God doth not continue in sin; for the Spirit of God remaineth in him; and he cannot continue in sin, because he is born of God, *having received that holy Spirit of promise*" (JST, 1 John 3:9). This verse suggests that spiritual rebirth at a certain level is associated also with receiving the Holy Spirit of Promise or having one's calling and election made sure (see, for example, D&C 124:124). It is not clear from the account in Mosiah whether this blessing was extended at this time to Benjamin's people.

They became, as mentioned above, candidates for the church of the Firstborn (see Mosiah 5:7 in connection with D&C 93:22).

The full reception of the Holy Ghost is the key to rebirth. Elder Bruce R. McConkie wrote: "Mere compliance with the formality of the ordinance of baptism does not mean that a person has been born again. No one can be born again without baptism, but the immersion in water and the laying on of hands to confer the Holy Ghost do not of themselves guarantee that a person has been or will be born again. The new birth takes place only for those who actually enjoy the gift or companionship of the Holy Ghost, only for those who are fully converted, who have given themselves without restraint to the Lord."[10]

These Nephites were "alive in Christ because they enjoy[ed] the companionship of the Spirit."[11] They were immersed in the Spirit, which they had received as a constant possession.

Speaking of permanent possession of the Spirit, Brigham Young taught that when one has been proved, and when one has labored and occupied himself sufficiently in obtaining the Spirit, if he would adhere to the Spirit of the Lord strictly, it should become in him a fountain of revelation.

"After a while the Lord will say to such, 'My son, you have been faithful, you have clung to good, and you love righteousness, and hate iniquity, from which you have turned away, and now you shall have the blessing of the Holy Spirit to lead you, and be your constant companion, from this time henceforth and forever.['] Then the Holy Spirit becomes your property, it is given to you for a profit, and an eternal blessing. It tends to addition, extension, and increase, to immortality and eternal lives."[12]

What is arresting here is that Benjamin's people were already commandment keepers. It is not a mighty change from evil to goodness that they have undergone, like Alma or Paul, but a transformation from basic goodness to something that exceeded their ability even to describe. This much they did say: "The Spirit of the Lord Omnipotent . . . has wrought a mighty change in us, or in our hearts, that we have no more disposition to do evil, but to do good continually" (Mosiah 5:2).

RECEIVING THE NAME

What, then, distinguished Benjamin's community "above all the people which the Lord God hath brought out of the land of Jerusalem" (Mosiah 1:11)? Perhaps this was the first time among all the people brought out from the land of Jerusalem that a king and a priest—in the tradition of Adam, Enoch, and Melchizedek—had succeeded in bringing his people to this point of transformation: he had caused them as a community actually to receive the name of Christ.

But what does it mean to receive the name of Christ? We remember that when we take the sacrament, we signify not that we have fully taken the name, but that we are willing to take the name (Moroni 4:3; D&C 20:77; compare Mosiah 5:5). Elder Dallin H. Oaks emphasized the word *willingness,* pointing to a future consummation:

"The Lord and his servants referred to the . . . *temple* as a house for 'the name' of the Lord God of Israel. . . . In the inspired dedicatory prayer of the Kirtland Temple, the Prophet Joseph Smith asked the Lord for a blessing upon 'thy people upon whom thy name shall be put in this house' (D&C 109:26.) . . . [B]y partaking of the sacrament we witness our willingness to participate in the sacred ordinances of the temple and to receive the highest blessings available through the name and by the authority of the Savior when he chooses to confer them upon us."[13]

Elder Bruce R. McConkie linked becoming a son or daughter of God with temple ordinances: "The ordinances that are performed in the temples are the ordinances of exaltation; they open the door to us to an inheritance of sonship; they open the door to us so that we may become sons and daughters, members of the household of God in eternity. . . . They open the door to becoming kings and priests and inheriting all things."[14]

In connection with being born again, Benjamin's people may have received something of a temple endowment. In fact, we find in Benjamin's discourse essential temple themes pertaining to the creation, fall, atonement, consecration, and covenant making. Benjamin's last words pertain to being "sealed" to Christ and receiving eternal life (Mosiah 5:15). Of course, important endowment elements are missing from the record, but had they been administered on this occasion, or at some later point, they would not, because of their sacred nature, have been included in our present Book of Mormon account. Nevertheless, King Benjamin's people received an endowment of spiritual knowledge and power which took them from being good people to being Christlike people—all in a temple setting. What they experienced through the power of the priesthood was a revelation of Christ's nature and the power to be assimilated to his image.

The account of Benjamin's people is compelling in its promise of that which awaits the diligent seeker of Christ. Ultimately, many spiritual questions are answered only after one's own personal experience, to which experience the Lord generously extends invitation. The Lord said on one occasion to a group of Saints, "Ye are not able to abide the presence of God now, neither the ministering of angels; wherefore, continue in patience until ye are perfected" (D&C 67:13). But he also taught: "Seek the face of the Lord always, that in patience ye may possess your souls, and ye shall have eternal life" (D&C 101:38). The message encourages diligence as well as patience for the fulfillment of the promise.

It is the privilege and responsibility of a community's priesthood leader, through faith and labor with his people, to bring them to a higher spiritual plane in their quest to return to God. Is it possible that Benjamin had been praying that the Lord would send his power to bring to pass a spiritually transforming experience for his people? The Lord sent his angel to declare to the king that power would be given to cause the people to be spiritually reborn, to become sons and daughters of Christ, and to receive the sacred name forever. It is the spiritual rebirth as a community and the taking of the name in a temple setting that distinguished them from those whom the Lord had previously brought out of Jerusalem (Mosiah 1:11). The people tasted the glory of God and came to a personal knowledge of him; through the power of the Holy Spirit they experienced the mighty change of heart and the mystery of spiritual rebirth.

Much of the Book of Mormon is devoted to that comprehensive change described in Benjamin's speech. That may be the reason President Ezra Taft Benson pled with us to feast on this book. He wrote, "When we awake and are born of God, a new day will break and Zion will be redeemed. May we be convinced that Jesus is the Christ, choose to follow Him, be changed for Him, captained by Him, consumed in Him, and born again."[15]

Notes

Revised from *King Benjamin's Speech, "That Ye May Learn Wisdom,"* ed. John W. Welch and Stephen D. Ricks (Provo: Foundation for Ancient Research and Mormon Studies, 1998), 277–94.

1. The emphasis in this verse and in subsequent scriptural passages in this chapter has been added by the author.
2. *Words of Joseph Smith*, 244; corrections and emphasis added.
3. Ibid., 38–39.
4. *Teachings of the Prophet Joseph Smith,* 159.
5. Ibid.
6. *Words of Joseph Smith*, 306, n. 30.
7. David B. Haight, "The Sacrament—and the Sacrifice," *Ensign,* November 1989, 60; emphasis added.
8. *Teachings of the Prophet Joseph Smith*, 297.
9. *Words of Joseph Smith,* 12; see also D&C 84:19–25; and JST, Exodus 34:1–2.
10. Bruce R. McConkie, *Mormon Doctrine*, 2nd ed. (Salt Lake City: Bookcraft, 1966), 101.

11. Bruce R. McConkie, *A New Witness for the Articles of Faith* (Salt Lake City: Deseret Book Co., 1985), 285.
12. Brigham Young, in *Journal of Discourses*, 2:135; emphasis added.
13. Dallin H. Oaks, "Taking upon Us the Name of Jesus Christ," *Ensign*, May 1985, 81; emphasis added.
14. Bruce R. McConkie, Conference Report, October 1955, 12–13; see also Revelation 3:12; 14:1.
15. Ezra Taft Benson, "Born of God," *Ensign*, July 1989, 5.

THE BROTHER OF JARED
AT THE VEIL

The temple is the narrow channel through which one must pass to reenter the Lord's presence. A mighty power draws us through that channel—the sealing power of the at-one-ment of the Lord Jesus Christ. The Savior's *at-one-ment* is another word for the sealing power. By the power of the at-one-ment, the Lord draws and seals his children to himself in the holy temples.

In scripture we can study how the ancient great ones were drawn through that narrow channel to find their heart's desire: we find, for example, Adam, cast out, bereft of his Lord's presence, searching relentlessly in the lonely world until he finds the keys to that passage to the Lord. Abraham searched for his priesthood privileges (Abraham 1:1–2) and after a diligent quest exclaimed, "Thy servant has sought thee earnestly; now I have found thee" (Abraham 2:12). Moses on Horeb, Lehi at the tree, Nephi on the mountaintop—all these men conducted that search which is outlined and empowered in the temple endowment, gradually increasing the hold, the seal, between themselves and their Lord.

This was the very search for which they were put on earth: to rend the veil of unbelief, to yield to the pull of the Savior's sealing power, to stand in the Lord's presence, encircled about in the arms of his love (D&C 6:20; 2 Nephi 1:15). This, then, is the temple endowment: having been cast out, to search diligently according to the revealed path, and at last to be clasped in the arms of Jesus (Mormon 5:11).

In particular, I wish to briefly focus in this chapter on some of the temple elements in the experience of the brother of Jared: (1) the Tower of Babel, (2) his period of probation, (3) his experience at the cloud-veil, and (4) some observations on faith and knowledge as revealed in the brother of Jared's search for the heavenly gift. One can see that these four elements follow a temple pattern: a false religion is offered; a period of probation or trial of faith is provided; and, upon obedience, light and knowledge are granted.

THE TOWER OF BABEL

The brother of Jared's rejection of the spiritual chaos at the Tower of Babel was a critical part of his ultimate endowment. By ancient tradition the Tower of Babel was inspired by Nimrod, a grandson of Ham, who sought to dethrone God by bringing men into constant dependence on his, Nimrod's, power. A multitude followed Nimrod, persuaded that it was cowardice to submit to God. The people began to build the tower, which was apparently some type of temple, as their objective was to reach heaven by means of the tower.[1] God's response was to break up their evil combination by scrambling their languages, thus depriving them of the powerful Adamic language.[2] The name *babel* means, in Akkadian, "gate of God" and is a play on the Hebrew *balal,* meaning "to mix or confound." It is apparent, then, that the Tower of Babel was a counterfeit gate of God, or temple, that Ham's priesthood-deprived descendants built in rebellion against God.

PROBATION

Jared and his family and friends rejected this temple and were spared the Lord's punishments. The Jaredite community enjoyed both the spirit of at-one-ment and the Adamic language and wanted to enlarge their privileges of righteousness, not diminish them. Thus they set out on the quest that is initiated by a period of stringent testing and training (gathering of animals and plants, trekking through wilderness, building two sets of barges, and enduring strong chastening). As their obedience and sacrifice increased, so did their privileges with the Lord, for "the Lord did go before them, and did talk with them as he stood in a cloud, and gave directions whither they should travel" (Ether 2:5). Successful navigation of their tests brought the brother of Jared to the need for more light and thus to the mount Shelem.

THE BROTHER OF JARED AT THE CLOUD-VEIL

The word *shelem* has three main Hebrew consonants forming a root word that spans a wide spectrum of meanings: peace, tranquility, contentment, safety, completeness, being sound, finished, full, or perfect. *Shelem* (and *shalom*) signify peace with God, especially in the covenant relationship. It also connotes submission to God, which we see in the Arabic words *muslim* and *islam*. In particular, *shelem* has reference to the peace offering of the law of sacrifice, which corresponds to the seeking of fellowship with God,[3] and thereby has a relationship to the meanings of the at-one-ment; that is, *shelem,* fellowship, sealing, and at-one-ment have an obvious relationship. When the brother of Jared carried the stones in his hands to the top of the mount, whether or not a temple peace offering is implied, he sought a closer fellowship or at-one-ment with the Lord. Therefore, the mount is called *shelem* "because of its exceeding height" (Ether 3:1), not because *shelem* means great height, but rather because it suggests a place that is suitably high for temple activity.

The small stones themselves suggest meanings beyond their practical use in the barges. Note that he did *molten* the stones, extracting them from the rock of the mount itself and shaping them by fire; white, clear, and glasslike, they evoke the Urim and Thummim (Hebrew words that mean "lights and perfections"). What is the relationship between these sixteen small stones and the two Urim and Thummim stones that the Lord gave the brother of Jared later on? It seems that the brother of Jared was led to fashion that which would give his community not only practical light but spiritual light as well; indeed, they were the very instrument of his calling as prophet, seer, and revelator. The small stones evoke the white stone mentioned in Revelation 2:17 and explained in Doctrine and Covenants 130:10–11, which stone becomes a Urim and Thummim to those who come into the celestial kingdom, "whereon is a new name written, which no man knoweth save he that receiveth it. The new name is the key word."

At the top of the mount, the brother of Jared seems to be operating under the influence of forces of which he is not fully conscious, but which his spirit seems to understand. He says that he is there for light, but his words reveal that his greatest concern is his unredeemed nature. He even appears to be afraid of the Lord's anger here and is so overcome with his inadequacy that he seems to be fighting the temptation to withdraw. It is with deliberate courage that he presses on past this fear, taking heart in the knowledge that the Lord has commanded him to ask and receive what he needs in spite of his fallen nature.

The fear he manifests suggests similar scenes in at least two other places in scripture when people have a close encounter with the Lord. The first example is King Benjamin's people, who fall to the earth "for the *fear* of the Lord had come upon them. And they had viewed themselves in their own carnal state. . . . And they all cried aloud with one voice, saying: O have mercy, and apply the atoning blood of Christ that we may receive forgiveness of our sins, and our hearts may be purified; for we believe in Jesus Christ" (Mosiah 4:1–2).[4] They experience pain and fear at their spiritually induced awareness of their fallenness in contrast to God's perfection. They plead for and receive a cleansing response from the Lord.

The second example comes from Isaiah's vision of the Lord. "Woe is me! for I am undone; because I am a man of unclean lips, and I dwell in the midst of a people of unclean lips: for mine eyes have seen the King, the Lord of hosts" (Isaiah 6:5). The Lord responds by cleansing him in his presence.

As the unredeemed soul, even a guiltless one, closes the gap between himself and his Maker, he perceives the contrast as so overwhelmingly great that he is sorely tempted to shrink back, to give up the quest. Those who will not be redeemed do shrink, overcome by fear of this encounter (consider the example of the Israelites in Exodus 20:18–21, which is discussed in chapter 4); but those who are determined to be redeemed press boldly on, and, exercising mighty faith, penetrate the veil and receive the transformation they so desire.

Standing now before this cloud-veil, having asked for light, the brother of Jared is stunned to see a finger appearing through the cloud-veil. He falls to the ground, struck with fear, because he *knows* what he sees. What he had held for so long in his "eye of faith" has just been visually confirmed. He has, to use Moroni's language, "ren[t] that veil of unbelief" (Ether 4:15) with his persistent believing-as-though-he-were-seeing, and has in some marvelous way operated the law that quickens and focuses his spiritual eyes. He had asked for the finger to touch the stones, and that is what he saw—what he asked for and believed. As Elder Boyd K. Packer observes, the world says, "'seeing is believing . . . show me!'" "When," he says, "will we learn that in spiritual things . . . *believing is seeing?* Spiritual belief precedes spiritual knowledge."[5]

The Lord says to the brother of Jared: "Because of thy *faith* thou hast seen; . . . for were it not so ye could not have seen my finger. Sawest thou more than this?" (Ether 3:9). It must have been with pounding heart that the brother of Jared said, "Nay; Lord, show thyself unto me" (v. 10). A further dialogue then takes place at the cloud-veil, the Lord testing the brother of Jared's desire and preparation, after which he says, "Ye are redeemed from the fall; therefore ye are brought back into my presence; therefore I show myself unto you" (v. 13). The brother of Jared receives the heavenly gift, described by Moroni in Ether 12: "For it was by faith that Christ showed himself unto our fathers, . . . and prepared a way that thereby others might be partakers of the heavenly gift. . . . Wherefore, ye may also have hope, and be partakers of the gift, if ye will but have faith. Behold it was by faith that they of old were called after the *holy order* of God. . . . Wherefore, he showed not himself until after their faith" (vv. 7–12). President Ezra Taft Benson explained the holy order of God: "To enter into the order of the Son of God is the equivalent today of entering into the fullness of the Melchizedek Priesthood, which is only received in the house of the Lord."[6]

FAITH AND KNOWLEDGE

The brother of Jared's experience dramatizes the difference between *faith* and *knowledge*. We can see that the brother of Jared did not have a perfect knowledge before he went through the veil because he expressed fear and surprise at what he saw and learned. The Lord says that it was not the brother of Jared's perfect knowledge that dissolved the veil; rather, it was his exceeding faith (Ether 3:6–9). It seems that Moroni means to say that once the brother of Jared had seen the Lord, he *then* had perfect knowledge of the Lord, and the Lord could not then withhold anything from him. Moroni says: "And after the brother of Jared had beheld the finger of the Lord, because of the promise which the brother of Jared had obtained by faith, the Lord could not withhold anything from his sight; wherefore he showed him all things, for he could no longer be kept without the veil" (Ether 12:21).

The knowledge given by the Holy Ghost, the first Comforter, is not a perfect knowledge, though it prepares and draws the seeker to that perfect knowledge. Faith, produced by the revelations of the Holy Ghost, is an assurance or *pre*-knowledge that what the Lord says is true (Alma 32:34). But faith is designed to proceed and become perfect knowledge, which is seeing something for ourselves after we have believed in, and been obedient to, the assurances of the Holy Ghost.

Faith is not an end in itself; it is a means to an end—and that end is to be like and to be with the Lord. When we say in our testimony meetings, "I know that the Lord Jesus lives," without having actually seen him, we mean that the Holy Ghost has given that assurance to our souls. But we do not have a perfect knowledge until, after an extended period of probation, we see for ourselves as the brother of Jared did. Joseph Smith observed, "Men at the present time testify of heaven and of hell, and have never *seen* either—and I will say that no man *knows* these things without this."⁷ Faith in the Lord Jesus Christ leads in one direction, and that is into the Lord's presence.

Moroni teaches this principle when he says, "And he [the brother of Jared] *saw* . . . and he had faith no longer, for he *knew,* nothing doubting" (Ether 3:19). A small sampling of several pertinent scriptures will show that the Lord often uses the word *know* with the word *see* when referring to spiritual knowledge.

1 Nephi 5:4: "If I had not *seen* the things of God in a vision I should not have *known* the goodness of God."

3 Nephi 11:15: "The multitude . . . did *see* with their eyes and did *feel* with their hands, and did *know* of a surety." [The Prophet Joseph said, "No one can truly say he knows God until he has handled something, and this can only be in the Holiest of Holies."⁸]

Alma 36:26: "Many have been born of God, and have tasted as I have tasted, and have *seen* eye to eye as I have *seen;* therefore they do *know* of these things . . . as I do *know*."

D&C 45:46: "You now *behold* me and *know* that I am."

D&C 50:45: "And the day cometh that you shall hear my voice and *see* me, and *know* that I am."

D&C 93:1: "Every soul who forsaketh his sins and cometh unto me, and calleth on my name, and obeyeth my voice, and keepeth my commandments, shall *see* my face and *know* that I am."

In a comment on 1 Peter 1:3–5, the *Lectures on Faith* make it clear that seeing the Lord is a pivotal point in the acquisition of knowledge. "[Peter] says that all things that pertain to life and godliness were given unto them through the *knowledge* of God and our Savior Jesus Christ. And if the question is asked, how were they to obtain the knowledge of God? (for there is a great difference between believing in God and knowing him. . . . And notice, that all things that pertain to life and godliness were given through the *knowledge of God*) the answer is given—through faith they were to obtain this knowledge; and, having power by faith to obtain the knowledge of God, *they could with it obtain all other things which pertain to life and godliness.*"[9]

Joseph Smith says similarly in another place: "The Lord will teach him [the receiver of the Second Comforter] face to face and he may have a *perfect knowledge* of the mysteries of the kingdom of God, and this is the state and place the ancient saints arrived at."[10] And the Prophet Joseph again: "Then knowledge through our Lord and Savior Jesus Christ is the grand key that unlocks the glories and mysteries of the kingdom of heaven."[11]

Joseph speaks of the kind of experience that the brother of Jared had and makes a connection to temple ordinances: "God hath not revealed anything to Joseph, but what He will make known unto the Twelve, and even the least Saint may know all things as fast as he is able to bear them, for the day must come when no man need say to his neighbor, Know ye the Lord; for all shall know Him . . . from the least to the greatest. How is this to be done? It is to be done by this *sealing power,* and the other Comforter spoken of, which will be manifest by revelation."[12]

Moroni says that "there never were greater things made manifest than those which were made manifest unto the brother of Jared" (Ether 4:4), but he says that they will not go forth to us, the Gentiles, until the day that we repent and become clean and sanctified and exercise faith like the brother of Jared. Then he says that the Lord will manifest unto the Gentiles the things the brother of Jared saw, even to the unfolding all his revelations (Ether 4:6–7):

"Come unto me, O ye Gentiles, and I will show unto you the greater things, the knowledge which is hid up because of unbelief. Come unto me, O ye house of Israel, and it shall be made manifest unto you how great things the Father hath laid up for you, from the foundation of the world; and it hath not come unto you, because of unbelief. Behold, when ye shall rend that veil of unbelief . . . then shall ye *know*" (Ether 4:13–15).

These possibilities pertain perhaps to this life, perhaps to the life to come, but the pattern of the brother of Jared points the way. Having rejected all counterfeit worship, having pushed on past all comfortable way stations, having sacrificed to come up to the full measure of obedience to the Lord, the brother of Jared received his endowment on the top of mount Shelem, where the Savior of the world sealed him his. President Benson taught: "God bless us to receive all the blessings revealed *by Elijah the prophet* so that our callings and election will be made sure.

"I testify with all my soul to the truth of this message and pray that the God of Abraham, Isaac, and Jacob will bless modern Israel with the compelling desire to seek all the blessings of the fathers in the House of our Heavenly Father."[13]

Notes

Revised from *Temples of the Ancient World*, ed. Donald W. Parry (Salt Lake City: Deseret Book Co., 1994), 388–98.

1. See Josephus, *Antiquities* (Grand Rapids, Mich.: Kregel Publications, 1994), I, 4.
2. See Joseph Fielding Smith, *The Way to Perfection* (Salt Lake City: Genealogical Society of Utah, 1935), 69.
3. See Francis Brown, S. R. Driver, Charles A. Briggs, *The New Brown, Driver, and Briggs Hebrew and English Lexicon of the Old Testament* (Boston: Houghton, Mifflin and Co., 1907), 1022–24; also LDS Bible Dictionary, s.v. "Sacrifices," 767.
4. The emphasis in this verse and in subsequent scriptural passages in this chapter has been added by the author.
5. Boyd K. Packer, "What Is Faith?" in *Faith* (Salt Lake City: Deseret Book Co., 1983), 43; emphasis added.
6. Ezra Taft Benson, "What I Hope You Will Teach Your Children about the Temple," *Ensign*, August 1985, 8.
7. *Words of Joseph Smith*, 10; spelling and punctuation have been standardized.
8. Ibid., 120.
9. *Lectures on Faith*, 7:18; emphasis added.
10. *Words of Joseph Smith*, 5; emphasis added; spelling and punctuation have been standardized.
11. *Teachings of the Prophet Joseph Smith*, 298.
12. Ibid., 149.
13. Benson, "What I Hope," 10; emphasis in original.

THE DYNAMICS OF APOSTASY: FROM MALACHI TO JOHN THE BAPTIST

Three distinct features characterize the revealed religion of the Old Testament: scripture, temple, and prophecy. Each of these suffered corruption in the apostasy that ended the Old Testament period and ripened in the intertestamental period (c. 500 B.C. to A.D. 30). We have no evidence that God sent any more prophets to Israel in the Old World after Malachi, at about 490 B.C. "The keys, the kingdom, the power, [and] the glory, had departed from the Jews," taught Joseph Smith, and "John, the son of Zachariah, by the holy anointing, and decree of heaven" would be the next legal administrator to hold the prophetic office.[1] Judaism, as it appeared in the intertestamental period, was not the revealed religion of the Old Testament. It had suffered mutation. The process followed this path: a steadily increasing rejection of living prophets (1200–500 B.C.), the cessation of revelation (c. 490 B.C.), and changes that developed as the result of the absence of revelation, such as reliance on tradition and the interpretations of men, tampering with scripture, and the changing of doctrine. The loss of divine gifts in Judah had implications not only for the Jews but for the Christians as well, who would travel the same path of apostasy within a very few years of John's and Jesus' restoration. Because apostasy follows a discernible pattern, our purpose here is to analyze the dynamics of apostasy in general but specifically to identify its manifestations in doctrine, scripture, and sources of revelation between Malachi and John the Baptist.[2]

THE NATURE OF APOSTASY

The purpose of the Church in any dispensation is to prepare the Saints by the power released through priesthood ordinances to acquire the divine nature and stand in God's presence. Jesus rebuked the Pharisees and scribes for thwarting that divine purpose: "Woe unto you, lawyers! for ye have taken away the key of knowledge: ye entered not in yourselves, and them that were entering in ye hindered" (Luke 11:52). The Joseph Smith Translation of this passage clarifies *knowledge* as "the fulness of the scriptures" (JST, Luke 11:53). Apostasy changes doctrines, deletes covenants, and tampers with scripture to justify the positions of the apostates.

Scripture reveals fundamental principles about the corruption of truth. The underlying dynamic of apostasy is pride, as we see from the Lord's reproach in Luke: "Woe unto you, Pharisees! for ye love the uppermost seats in the synagogues, and greetings in the markets" (Luke 11:43). A scant seventy years later the apostle John would describe the apostasy developing in the Church of Jesus Christ: "I wrote unto the church: but Diotrephes, who loveth to have the preeminence among them, receiveth us not . . . prating against us with malicious words: and not content therewith, neither doth he himself receive the brethren, and forbiddeth them that would, and casteth them out of the church" (3 John 1:9–10).

Apostasy, by its nature, takes place *within* the covenant people: "They [antichrists] went out from us, but they were not of us" (1 John 2:19). The word *apostasy* comes from the Greek word *apostasia,* which means "rebellion." Satan was the first apostate and the father of apostasy. He revealed the essential motive behind apostasy when he challenged God: "Wherefore give me thine honor" (Moses 4:1). Indeed, apostasy is a fruit of pride and is characterized by hatred. President Ezra Taft Benson spoke with penetrating insight on pride:

"The central feature of pride is enmity—enmity toward God and enmity toward our fellowmen. *Enmity* means 'hatred toward, hostility to, or a state of opposition.' It is the power by which Satan wishes to reign over us."[3]

Jesus highlighted the malice of apostates as he charged the Pharisees with the blood of all the martyred prophets and Saints shed from the foundation of the world (Luke 11:50–51). Luke recorded the Pharisees' hatred of Jesus, as they were "laying wait for him, and seeking to catch something out of his mouth, that they might accuse him" (Luke 11:54). These Pharisees who sought Jesus' life were the descendants of that apostasy that thrived in Old Testament times and into the succeeding period.

As apostasy ripens, it develops from individual and random wickedness to an increasingly sophisticated state of organization. The apostasy that brought prophecy to an end and later flourished in the intertestamental period probably began to assume an organizational structure in the days of Ezra (c. 458 B.C.), who was a priest and scribe but also a political appointee of the Persian emperor, Artaxerxes I (Ezra 7:12–26). Ezra's objective was "to teach in Israel statutes and judgments" (Ezra 7:10). It was apparently this time that marked the beginning of the institution of the synagogue—the gathering place for instruction in the Law. The scribal tradition seems to have taken shape in this period also. Some of the educated Jews began to use knowledge of the scriptures as a means of personal promotion (e.g., Matthew 23:1–12). The scribes, professional scholars of the law, became the most prominent citizens of the community. From these beginnings rose the institutions of the Pharisees, scribes, and lawyers of Jesus' day. Along with the Sadducees—the priests and their allies—these groups were responsible for the dissemination of apostate ideas.

As awareness increased that God no longer worked among men, the Jews' spiritual leaders began to focus on the written word as if it were their messiah and vested it with the power of salvation.[4] Nephi wrote that to his people the Law was dead, that they only kept it because they were commanded to (2 Nephi 25:25). But he then foreshadowed the difficulty that the Jews would have in letting go of the Law: the Jews "need not harden their hearts" against Christ "when the law ought to be done away" (2 Nephi 25:27). Later, Abinadi found that the priests of Noah did indeed believe that salvation came not by Christ but by the law of Moses (Mosiah 12:31–32).

Among the Jews, then, prophecy had ceased, the canon of scripture was closed, and private interpretation of scripture supplanted divine revelation. Three things happened as a wave of darkness and confusion washed over the Jewish community: divine revelation withdrew; revealed doctrines began to change, especially the doctrine of God; and scriptural mutations appeared.

THE TRANSFORMATION OF SCRIPTURE

That scripture tampering is a fundamental component of apostasy is revealed by the Book of Mormon. Nephi prophesied that evil men would deliberately take key passages from the prophets' records "that they might pervert the right ways of the Lord" (1 Nephi 13:23, 25–27, 29). Moses, too, learned from the Lord that men would esteem his words as naught and take many of them from the book that Moses would write (Moses 1:41). Just when and how these changes happened is not clear, but that they did happen is supported by documentary evidence.

The Old Testament describes Ezra as "a ready scribe in the law [Torah][5] of Moses" (Ezra 7:6) who had "prepared his heart to seek [search, study, or interpret] the law of the Lord, and to do it" (Ezra 7:10). Not only did he "read in the book in the law of God distinctly," he "gave the sense, and caused them to understand the reading" (Nehemiah 8:8). This practice of interpreting the law is what the scribes strove to do. The scribes were the lawyer class of Jewish life, the "doctors of the law." Educated intellectuals, trained in the practical or civil aspects of the law of Moses, the scribes interpreted and applied the law for everyday life.[6] In the absence of divine revelation, the law became susceptible to fresh interpretations for each succeeding generation. To make applications to every aspect of life, the scribes developed a running commentary on the words of scripture. This method of teaching was known as Midrash (from Hebrew *drs,* meaning "to search," with the implication of searching out the true interpretation). In places where the law was unclear, its meaning had to be interpreted and expounded. The scribes and later the rabbis (teachers who came after and continued the tradition of the scribes) built "a fence around the Law,"[7] which consisted of cautionary rules, such as that forbidding not simply the use but even the handling of tools on the Sabbath day. Thus by this "fence," a man would be halted before he found himself in danger of breaching the law of God. These rules and interpretations of the Law constituted the "oral tradition." Therefore the law was seen to have two parts—written and oral—and each was of equal authority. Each was also viewed to be of equal antiquity, for Moses himself, so they believed, had received the law, written and oral, at Sinai, from which it had been handed down through successive generations of faithful men. It was in part a conflict over the authority of the oral tradition that led to the appearance of the two major parties of Jesus' day: the Pharisees and the Sadducees. The Pharisees were the inheritors, guardians, and perpetuators of the oral tradition; the Sadducees regarded the written law alone as authoritative.[8]

As time went on, the oral tradition became disassociated from the text of the written scriptures and no longer needed to find its justification there. The oral tradition took on a life of its own. The Mishnah (from verbal teaching by repeated recitation; Hebrew root, *snh*) is the collection of the rabbis' interpretations, or the oral tradition, written down about A.D. 200. The Talmud (literally, "learning") is a compilation consisting of the Mishnah, together with the subsequent discussions and commentaries (the *Gemara,* literally, "completion") which arose in the Jewish schools.[9] The Jews' scriptural tradition had become a mixing of scripture with the philosophies of men. They sought guidance from this source rather than from God. The inspired person sees scripture as a catalyst to revelation; apostasy founders in literalism.

From the time of Ezra onward, the Judaism that gradually developed attached great importance to the law; the other sections of the scripture—the Prophets and the Writings—had less importance. And during the period 500–200 B.C., another gradual shifting of emphasis occurred from the temple to the law. By about 200 B.C., the religion of the law was solidly established.[10] The attitude toward the structure of the Hebrew Bible itself reflects the Jewish consciousness of the ebbing away of spiritual power in Israel: the rabbis saw Moses' law as the most sacred, the Prophets next, and the Writings they saw as least sacred.[11] Because God had spoken face to face with Moses, the law was the holiest writing, in their view. The prophets received their revelation through the Holy Spirit, not directly from God; and the Writings came from individuals of lesser prophetic gifts. Thus the golden age of revelation in the mind of the Jews had been the age of Moses, and each succeeding age had diminished in splendor.

By about 250 B.C., many Jews who lived scattered throughout the Mediterranean area (in the Diaspora, or Dispersion) had lost a knowledge of Hebrew, the language of the Old Testament. Therefore, a Greek translation of the Hebrew scriptures was commissioned, called the Septuagint (often abbreviated as LXX). But with the translation of revealed religion into the Greek idiom, many Greek religious concepts slipped into the religion of the Jews.

THE CORRUPTION OF THE DOCTRINE OF GOD

Hellenism, the Greek culture that grew to dominate the Near East after Alexander the Great's conquests of the fourth century B.C., had a profound effect on Jewish thought. One of the doctrines most affected by the hellenization of Judaism was the doctrine of the nature of God. This doctrine is one of the first to undergo modification in any apostasy. The object is generally to deprive God of his body and his feelings and thus to render him amorphous and distant. With these transformations, true religion loses the doctrine of the deification of men and the accompanying commitment to fidelity in marriage and to stable, nurturing families.[12] This loss of the knowledge of the true God happened early in Christianity as well. Alma found it among apostate Zoramite Nephites (Alma 31:15). The first knowledge restored to Joseph Smith in the First Vision was that God the Father and his Son Jesus Christ have separate bodies in the form of men, a doctrine that had been officially absent from Christianity since the Council of Nicaea declared the consubstantiality of the Father and the Son (in A.D. 325). Hellenism and other forces influenced the shape and nature of God in apostate Judaism.

After Alexander's conquests (c. 300 B.C.), the Jews were surrounded by Greek culture and civilization. Particularly in the Diaspora (the Jewish colonies outside Palestine), many of the Jews who adopted the Greek language were deeply influenced by their hellenistic environment. With the absorption of Greek language and culture, Jewish thinking began to take on an increasingly Greek tone. A proliferation of apocryphal literature appeared in this period, manifesting a mixture of Jewish and Greek thought. The hellenized Jew Philo, a contemporary of Jesus and Paul, embraced Greek ideas of an utterly transcendent God who had nothing to do with the material world because of its inferior nature and had even caused the world to be created by inferior beings.[13] These ideas would within a few years help to produce sexual asceticism and docetism (the doctrine that Christ only *seemed* to be flesh) in the earliest Christian church. Hellenization was an aggressive movement, and the pressure exerted by sophisticated Greek philosophical systems was more than many Jews could resist. Again pride was at work as Jews became ashamed of the plain, revealed religion of their Israelite forefathers (e.g., 1 Maccabees 1:11–15, which documents the "uncircumcising" of Jews).

In contrast, the doctrine of God in the Old Testament was simple. God had not simply wound up the world like a clock and retired. The revealed God of the Old Testament sought constant interaction with his children. He spoke in terms of bride and bridegroom when he expressed the loving and nourishing covenant relationship that he desired with Israel: "I will betroth thee unto me for ever; yea, I will betroth thee unto me in righteousness, and in judgment, and in lovingkindness, and in mercies. I will even betroth thee unto me in faithfulness: and thou shalt know the Lord" (Hosea 2:19–20).

The Old Testament abounds in anthropomorphic expressions (i.e., expressions pertaining to human form) referring to God. The parts of the human body frequently serve as descriptions of the acts of God. "The eyes of the Lord" and the "ears of the Lord" occur in the Prophets and in Psalms; "the mouth of the Lord" speaks both in the Law and the Prophets; the heavens are the work of his fingers (Psalm 8:3), and the tablets of the covenant are written by the finger of God (Exodus 31:18). We read of his "nose," that is, the wrath of God; "his countenance" which he causes to shine, or, alternatively, which he hides. We read of "his right hand," "his arm," and "his sword." In Genesis he walks about in the garden (Genesis 3:8); he "goes down" in order to see what is being done on the earth (Genesis 11:5; 18:21) or to reveal himself there (Exodus 19:18; 34:5), and he goes up again (Genesis 17:22; 35:13). He sits on a throne (Isaiah 6:1) and causes his voice to be heard among the cherubim who hover over the ark of the tabernacle (Numbers 7:89). The hair of his head is as wool (Daniel 7:9), and Moses sees "his back" (Exodus 33:23).[14]

Anthropopathisms (humanlike emotions) abound as well: God expresses love, joy and delight, regret and sadness, pity and compassion, disgust, anger, and other feelings. One scholar observed that the early Israelites preferred making God immediate rather than distant:

"Ultimately, every religious expression is caught in the dilemma between, on the one hand, the theological desire to emphasize the absolute and transcendental nature of the Divine, thereby relinquishing its vitality and immediate reality and relevance, and on the other hand, the religious need to conceive of the Deity and man's contact with Him in some vital and meaningful way. Jewish tradition has usually shown preference for the second tendency, and there is a marked readiness to speak of God in a very concrete and vital manner and not to recoil from the dangers involved in the use of apparent anthropomorphisms."[15]

In time, however, the more sophisticated people did recoil at an anthropomorphic God. At least by post–Old Testament times, the

scribes and the rabbis who gave us the Mishnah found the anthropomorphisms offensive and took steps to make small textual changes
which they described as "biblical modifications of expression."[16] In
place of "I will dwell in your midst" (spoken by God) was substituted
"I shall cause you to dwell."[17] To avoid an objectionable anthropomorphism, the text of Exodus 34:24 may have been changed from
"to see the face of the Lord" *(lir' ot' et-pene yhwh)* to "to appear
before the Lord" *(lera' ot' et-pene yhwh).*[18] Early translations of scripture also changed such phrases as "he saw" or "he knew," referring to
God, to "it was revealed before him." "He went down" became "he
revealed himself"; "he heard" became "it was heard before him," and
so forth.

The translators of the Septuagint went further than the scribes in
spiritualizing the anthropomorphic or anthropopathic phrases of the
Bible. The "image of God" became in the Septuagint "the glory of
the Lord"; "the mouth of God" became "the voice of the Lord." They
deprived God of human emotions in careful, nonhuman repre-
sentations of God's wrath and pity.[19] Although the Israelites of
Moses' time could accept an anthropomorphic God, the Jews of
Jesus' time were ready to stone him for his proclamation that he was
Jehovah (John 8:58–59), at least in part because they had long sought
to obscure the doctrine of a God in human form.

Another means of distancing God was avoiding the use of his
name. By the third century B.C., saying the name *Yahweh* (Jehovah)
was avoided[20] and ultimately became illegal and blasphemous;
Adonai, translated "the Lord," was substituted for it. But God had
ordained his name as a key word by which a covenant person could
gain access to him (Moses 5:8; 1 Kings 8:28–29). Possession of the
name through priesthood authority meant possession of priesthood
power (Abraham 1:18–19). To forbid the name was to forbid access
to the power to acquire the divine nature and thus to prevent access to
God himself.[21]

Distancing God by making him amorphous and transcendent
makes him less demanding. If the Jews wanted to place distance
between them and God, they could do so, at least for the time being.

Jacob taught in the Book of Mormon that the Jews "despised the words of plainness, and killed the prophets, and sought for things that they could not understand.²² . . . For God hath taken away his plainness from them, and delivered unto them many things which they cannot understand, *because they desired it*. And because they desired it God hath done it, that they may stumble" (Jacob 4:14; emphasis added).

Perhaps it is helpful to note here that in any apostasy there are always innocent and well-intentioned victims. Not all promoters of false ideas have malignant intent; most are to some extent the victims of those who have gone before. But the Lord has made it apparent that the initiators of apostasy set out to destroy the work of the Lord (1 Nephi 13:27) and that many subsequent perpetrators of apostasy had the same objective. Joseph Smith wrote that many errors in the Bible were done in ignorance or by accident, but that many were the work of "designing and corrupt priests."²³

THE CESSATION OF DIVINE REVELATION

Malachi (c. 490 B.C.) wrote the last prophetic book in the Old Testament. After his time, the Jews clearly understood that prophecy had ceased. Even in Israel's earlier wicked days, God had continued to strive with them through prophets. But for the word of God to cease altogether among men reflects abysmal spiritual conditions.

Later Jews looked back on this part of Israel's history and acknowledged that the spirit of prophecy and revelation had been withdrawn. In medieval rabbinic literature, three Hebrew terms were used to represent God's revelation of himself to his children: *Ru'ah ha-Qodesh* (the Holy Spirit), *Shekhinah* (God's presence), and *Bath-qol* (a small voice). According to the rabbis, these manifestations thrived in Israel prior to the destruction of Solomon's temple (587 B.C.); but later they had been withdrawn. The second temple, rebuilt by Zerubbabel, was viewed as inferior to the first on at least five points: the absence of (1) the ark of the covenant, (2) the divine fire, (3) the

Urim and Thummim, (4) the Shekhinah (God's presence), and (5) the Holy Spirit.[24] The rabbis, finding themselves in a spiritual void, looked longingly back on the golden age of revelation of God to man and forward to the great messianic age, the "end of days," the culmination of which would be the restoration of the powers of the Holy Spirit.

Shekhinah is a rabbinic term that looks back on Old Testament times and refers to the presence of God's glory at such historic manifestations as indicated in Exodus 33:18, 22; Deuteronomy 5:22, 24; and 1 Kings 8:10–13. This majestic presence of God descended to "dwell" among men.[25] Israel constructed the tabernacle in order that this divine presence might dwell on earth and enter the Holy of Holies. The *Shekhinah* was often identified in rabbinic literature with the Holy Spirit and was frequently depicted in the form of a dove.[26]

The *Bath-qol,* translated "daughter of a voice" (implying a small voice), proclaimed God's will. In the Aramaic versions of the Bible, in the Midrash, and in the Talmud, the heavenly revelation is usually introduced with "a voice fell from heaven," or a voice "came from heaven," "was heard," or "proceeded from heaven." But the rabbis made a distinction between the *Bath-qol* and the Holy Spirit: whereas the prophets seemed to *possess* the Holy Spirit, the *Bath-qol* had no personal or persisting relationship with those who heard it. It could not be possessed.[27]

Even though a lesser gift than prophecy, the *Bath-qol,* according to the rabbis, had coexisted with prophecy in earlier times when the Spirit visited Israel. All three of these spiritual manifestations—the Holy Spirit, the *Shekhinah,* and the *Bath-qol*—were connected with the sign of the dove in rabbinic literature.

In Jewish tradition, three main views date the cessation of these manifestations of the Spirit and God's presence: (1) some date it as late as the destruction of Herod's temple (A.D. 70); (2) some date it from the destruction of the first temple (587 B.C.) when the Jews realized that the *Shekhinah* and the Holy Spirit of prophecy did not dwell in the second temple; (3) others point to the passing of the last three prophets (c. 500 B.C.). The Mishnah has references to the ceasing of divine revelation: "When the last of the prophets, Haggai, Zechariah, and Malachi, died, the Holy Spirit departed from Israel; nevertheless they made use of the Bath-qol."[28] Another passage says, "At first, before Israel sinned against morality, the Shekhinah abode with each individual; as it is said, 'For the Lord thy God walketh in the midst of thy camp' (Deuteronomy 23:14). When they sinned, the Shekhinah departed from them."[29] The writers of the Talmud believed that the spirit of prophecy vanished with the last prophets, and they acknowledged that thereafter the *Bath-qol* served instead. First Maccabees records, "Thus there was great trouble in Israel, such as had not been since the time when a prophet last appeared among them" (9:27). Another teacher declared, "In this world there is neither a prophet nor a Holy Spirit [Psalm 74:9], and even the *Shekhinah* has vanished on account of our sins [Isaiah 59:2]; but in the future world a new revelation will be vouchsafed to them."[30]

It was at the baptism of Jesus Christ that the signal appeared that the restoration of the spirit of prophecy had taken place. The Holy Spirit, descending like a dove, proclaimed the reconnection of heaven and earth (John 1:32–33). The dove—the ancient sign of the Holy Spirit, the Presence, and the small voice—was manifested to Israel. The medium was the message, and the message could not have been lost on a Jewish audience: here was clear evidence that the heavens were open again.

CONCLUSION

The apostasy of the Jews went right on, largely uninterrupted by the coming of Jesus Christ. After Jesus' and John's short ministries, those hellenistic influences that had produced the Jewish apostasy infiltrated the young Christian church as well; the same gospel mutations appeared in a remarkably short time, as we can see in the letters of Paul (e.g., Titus 1:14). The doctrine of God suffered corruption almost immediately at the hands of hellenized Christians, as did the scriptures (1 Nephi 13:24–29). The temple was destroyed again in

A.D. 70 as Jesus prophesied it would be (Matthew 24), and the spirit of prophecy had again flown by the end of the first century A.D.

Notes

Revised from *Studies in Scripture, Volume 4: 1 Kings to Malachi*, ed. Kent P. Jackson (Salt Lake City: Deseret Book Co., 1993), 471–83.

1. Sermon of 29 Jan. 1843, in *Times and Seasons* 4 (15 May 1843): 200.
2. This chapter deals with theological developments between the Old and the New Testaments. For a discussion of historical developments and circumstances during the same period, see Stephen E. Robinson, "The Setting of the Gospels," in *Studies in Scripture, Volume Five: The Gospels,* ed. Kent P. Jackson and Robert L. Millet (Salt Lake City: Deseret Book Co., 1986), 10–37.
3. Ezra Taft Benson, "Beware of Pride," *Ensign,* May 1989, 4; see also C. S. Lewis, *Mere Christianity* (New York: Macmillan, 1960), 111.
4. John 5:39–40: "You search the scriptures because you *think* that in them you have eternal life; and [yet] it is they that testify on my behalf. Yet you refuse to come to me to have life" (New Revised Standard Version; emphasis added).
5. Hebrew law or teaching; the Jews used the term to refer both to the law that God gave Moses on Sinai and to the first five books in the Old Testament, Genesis to Deuteronomy. These five books are also known as the Pentateuch.
6. Robinson, "The Setting of the Gospels," 10–37.
7. Herbert Danby, ed., *The Mishnah* (London: Oxford, 1983), 446; Aboth 1.1.
8. Robinson, "The Setting of the Gospels," 22–25, 29.
9. D. S. Russell, *Between the Testaments* (London: SCM, 1963), 60.
10. Ibid., 68.
11. For a listing of the books in the order in which they appear in the Hebrew Bible, see Kent P. Jackson, "God's Testament to Ancient Israel," *Studies in Scripture, Volume Three: Genesis to 2 Samuel* (Salt Lake City: Randall Book, 1985), 7.
12. For a full discussion of the effects of sexual asceticism on the doctrines of the deification of man, the nature of God, and the Atonement, see M. Catherine Thomas, "The Influence of Asceticism on the Rise of Text, Doctrine, and Practice in the First Two Christian Centuries," unpublished dissertation, Brigham Young University, 1989.
13. Philo, Conf. 179; *The Confusion of Tongues,* Loeb Classical Library 4 (Cambridge, Mass.: Harvard, 1930): 109.
14. "Anthropomorphism," *Encyclopaedia Judaica,* 17 vols. (Jerusalem: Keter, 1982), 3:53.
15. Ibid.
16. Ibid.
17. Ibid.
18. Ibid.
19. See also Bruce R. McConkie, *A New Witness for the Articles of Faith* (Salt Lake City: Deseret Book Co., 1985), 403.
20. "God, Name of," *Encyclopaedia Judaica,* 7:680.
21. For further discussion of the name of God, see Dallin H. Oaks, Conference Report, April 1985, 103.
22. That is, the Jews preferred the mysterious doctrines that they had created over plain doctrine. The Christians would do the same within a few years of Jesus' ascension.

23. The word *priests* presumably refers to any kind of religious leaders; see *Teachings of the Prophet Joseph Smith*, 327.
24. *Babylonian Talmud*, Yoma 21b.
25. "Shekhinah," *The Jewish Encyclopedia*, 12 vols. (New York: Funk and Wagnalls, 1901–6), 11:258.
26. Ibid., 11:260.
27. "Bath-qol," *Jewish Encyclopedia*, 2:589.
28. Yoma 9b and Sotah 48b.
29. Sotah 3b.
30. Arthur Marmorstein, *Studies in Jewish Theology* (Freeport, N.Y.: Books for Libraries Press, 1972), 125.

VISIONS OF CHRIST IN THE SPIRIT WORLD AND THE DEAD REDEEMED

A scene from Dante's *Inferno* will set the stage for this chapter. The poet Virgil brings Dante to the ledge of hell, called Limbo, and leads him into the dark world below. In the dim and tremulous air, the sound of sighing rises from the numberless congregation of souls suspended forever in sadness. Dante asks, thinking of the countless souls worthy of deliverance from this place, whether any-one by his own merit or that of another had ever gone from Limbo to a place of blessedness.[1]

Virgil answers that a Mighty One did come and took with him the ancient patriarchs and other righteous souls. But those who remained endured in hell without hope of deliverance.

Dante's theology of the plight of the unredeemed dead stands in stark contrast to the belief of the early Christians and those who remembered the teachings of the apostles. In the early Church there was no more prominent and popular belief than the Savior's descent to hell and the redemption of the dead. Ignatius (A.D. 35–107),[2] Polycarp (A.D. 69–155),[3] Justin Martyr (A.D. 100–165),[4] Irenaeus (A.D. 130–200),[5] Tertullian (A.D. 160–220),[6] and other early Christian writers either explicitly mention or allude to the Descent, several linking the Descent to redemptive work for the dead.

The word *hell* is a translation of *sheol* in Hebrew and *hades* in

Greek. Both terms refer to the place of departed spirits and not necessarily to that place where only the wicked go. Selwyn writes:

"The concept of Sheol or Hades as the abode of the dead generally, without ethical or other distinctions, was later differentiated to admit of distinct regions for the righteous and wicked respectively, as in *Enoch xxii.* As late as the *Psalms of Solomon* (cf. xiv. 6, xv. 11, xvi. 2) Hades was used for the place of punishment of the wicked, which is normally termed in N.T. Gehenna. The abode of the righteous, on the other hand, when a special term other than Hades is used, is spoken of as 'Abraham's bosom' (Luke xvi. 23) or 'Paradise' (Luke xxiii. 43.)"[7]

Although the belief in the Lord's descent into hell was firmly embedded in early Christian literature, the Bible provides little more than circumstantial evidence for the Descent. But the ancients believed that the spirit world lay under the earth, thus the idea of having to descend to arrive there.[8] Paul writes of the Savior's descending into the lower parts of the earth (Ephesians 4:9), although he may be referring not to the spirit world but simply to the Lord's birth on the earth. But Peter is probably the only one who makes explicit reference to the Lord's ministry in the spirit, and he does not use a term for *descent,* but rather a form of the verb *to go.* Nevertheless, I will use the term Descent hereafter to mean the Savior's journey to the spirit world, since that is the term the early Christians used. (The common Latin term for the Descent is *descensus ad inferos.*)

In modern times the Descent is perhaps the most neglected doctrine of early Christian theology. According to F. Loofs, a prominent biblical scholar, "the Decensus belongs in fact to a group of primitive Christian conceptions which are inseparable from views then current, but now abandoned, and which accordingly can now be appraised only in a historical sense, i.e. as expressions of Christian beliefs which, while adequate enough for their time, have at length become obsolete. . . . The modern mind cannot bring to it more than interest; we cannot now accept it as part of our faith. . . . It were fitting, therefore, that the Churches distinguished as Evangelical should omit the Article 'descendit ad inferos' from their programmes of instruction in Christian doctrine and worship."[9]

Indeed, what shall the world do with the doctrine of the Descent and the related doctrine of baptism for the dead? In a discussion of baptism for the dead, the New Testament scholar Edgar Goodspeed was asked, "Do you think it [baptism for the dead] should be practiced today?" He answered, "This is the reason why we do not practice it today. We do not know enough about it. If we did, we would practice it."[10]

My purpose will be to treat five major points on the Descent and the redemption of the dead as outlined in Doctrine and Covenants 138, which is President Joseph F. Smith's "Vision of the Redemption of the Dead." I will present supporting material from early Christian and Jewish writers to illustrate that the remarkable teachings pertaining to the redemption of the dead in section 138 represent truths accepted by the early Christians. I will comment little on Doctrine and Covenants 138 itself, but rather will use its major points relating to the Descent as an outline for my presentation. These five points are: (1) some history, translation, and interpretation of the much-disputed passages in 1 Peter which section 138 includes, i.e., 3:19 and 4:6, focusing on the baptismal context of these passages; (2) early Christian and Jewish evidence bearing on the Savior's redemptive work in the Descent; (3) evidence of belief in a division of spirits in the spirit world and belief in the anticipation by the righteous of the Messiah's appearance to them; (4) evidence of the organization of the righteous to take the gospel to the wicked; and (5) evidence of vicarious work for the dead in early Christianity.

First Peter 3:18–21 and 4:6 — History

The history of the interpretation of 1 Peter 3:19 (Christ preaching to the spirits in prison) and its context reflects the confusion that came in the wake of the Apostasy. I will touch here only lightly on the creative means by which exegetes have through the centuries wrested this passage. For greater detail, the reader may consult several writers who have made thorough studies of the complex factors, grammatical and exegetical, in the 1 Peter passages.[11] The difficulty does not actually lie in the passages, but in the minds of the interpreters who find a conflict here with their own views of the afterlife and the impossibility of progress or redemption there. Nearly all the interpretations of 1 Peter 3:19 and 4:6, from ancient times to our own day, are confined to the following possible interpretations: (1) The Lord preached the gospel to these spirits and offered them repentance. Under the influence of later theological ideas many commentators have been unwilling to admit this interpretation, maintaining instead (2) that Christ must have preached to them not hope but condemnation; or (3) that he preached only to those who were righteous; or (4) that he preached only to those who, though disobedient, repented in the hour of death; or (5) that he preached the gospel to those who had been just, and condemnation to those who had disobeyed.[12] The ancient Alexandrian Christian school of theology accepted the plain interpretation of 1 Peter 3:19 that Christ preached the gospel of hope to the unbelievers in the spirit world.

Later many commentators of the Middle Ages (as well as of modern times) did not find in this verse any allusion to the Descent, i.e., a ministry in the spirit world, at all. Where some did find a Descent, they interpreted it to mean deliverance of the Old Testament Saints only and the defeat of Satan in that event, which came to be known as the Harrowing of Hell; the belief that Christ had liberated any others than those holy persons became heretical.[13] Modern Catholic theology mostly tends to regard those who heard Jesus preaching in the spirit prison as sinners who converted before they died.[14] It appears that as time went on, almost no one understood the passages to mean that Jesus offered redemption to believers and unbelievers, saints and sinners; some restriction was imposed on the Lord's work.

We should note here that mention of the Descent appears in many early Christian creeds, such as the famous Apostles' Creed (A.D. 390). But the earliest creed of which we have record is known as the Fourth Formula of Sirmium (a council of Western bishops), which came thirty-one years earlier in A.D. 359 and was a descendant of many former Christian creeds that did not contain any mention of the Descent. I quote the section on the Descent from the Sirmium Creed: "And [Christ] descended to hell, and regulated things there, whom the gatekeepers of hell saw and shuddered, and [he] rose again."

After years of creeds without mention of the Descent, why was the Descent inserted into this creed? J. N. D. Kelly points out that this Sirmium Creed was drafted by a Syrian, Mark Arethusa, and that the Descent had a place very early in Eastern creed material.[15] He gives the Syrian *Didascalia* (a collection of miscellaneous precepts of professedly apostolic origin, c. A.D. 250) as example. There it says: "[Christ] was crucified under Pontius Pilate and departed into peace, in order to preach to Abraham, Isaac, and Jacob and all the Saints concerning the ending of the world and the resurrection of the dead."[16]

Why the Descent was so long neglected in the Western creeds is not clear, but Kelly speculates that Mark of Arethusa, having credal materials before him which contained mention of the Descent, felt the interpolation in the Western creed was important to show the full scale of the Savior's work of redemption.[17] At about the same time, synods at Nike (A.D. 359) and at Constantinople (A.D. 360) published creeds with a Descent clause. Rufinus (c. A.D. 404) records that the Aquileian creed contained the Descent clause, which he connected with 1 Peter 3:19, and he says that the Descent passage is included in that creed to explain "what Christ accomplished in the underworld."[18]

Even though many references to the Descent appear in early Christian literature, curiously, interpretations, reflections, or quotations of 1 Peter 3:19 are missing in the earlier Christian literature. No writer before Hippolytus (c. A.D. 200), Clement of Alexandria (A.D. 150–215), and Origen (A.D. 185–253) (who do indeed make clear connection between the Descent and 1 Peter 3:19) appears to allude to this verse.[19] The reason may lie in the difficulties mentioned in the connection of the Descent with redemption in 1 Peter 3:19.[20]

TRANSLATION, INTERPRETATION, AND BAPTISMAL CONTEXT OF 1 PETER 3:18–21 AND 4:6

"For Christ also hath once suffered for sins, the just for the unjust, that he might bring us to God, being put to death in the flesh, but quickened [the Greek may be rendered, 'made alive in'] by the Spirit: By which also he went and preached unto the spirits in prison" (1 Peter 3:18–19).

Much debate has taken place over what condition "alive in the spirit" implies. The King James Version gives "quickened in the Spirit" and capitalizes *Spirit,* implying the interpretation that the Lord went in the power of the Holy Spirit—that is to say, that this journey refers not to a descent but to his resurrection and to a proclamation of his triumph over the powers of evil.[21] But it is more likely that the two datives (flesh and spirit) should be understood as antithetical, meaning that when Christ's flesh was dead, his spirit continued alive into the spirit world. Joseph Smith taught this sense: "Now all those [who] die in the faith go to the prison of spirits to preach to the dead in body, but they are *alive in the spirit,* and those spirits preach to the spirits that they may live according to God in the spirit. And men do minister for them in the flesh, and angels bear the glad tidings to the spirits, and they are made happy by these means."[22]

Origen understood this sense of the verse that Jesus went in his spirit: "We assert that Jesus not only converted no small number of persons while he was in the body . . . but also, that when he became a spirit, without the covering of the body, he dwelt among those spirits which were without bodily covering, converting such of them as were willing to Himself."[23] Hippolytus (c. A.D. 155–236) wrote, "The Only-begotten entered [the world of spirits] as Soul among souls."[24]

For Augustine (A.D. 354–430), Bede (A.D. 673–735), Aquinas (c. A.D. 1225–74), and others, the difficulties in accepting the plain sense of 1 Peter 3:19 were insuperable. Although at first Augustine accepted the literal view of 1 Peter 3:19,[25] he later proposed a new interpretation, which was that Christ was *in* Noah when Noah preached repentance to the people of his time, and the spirits in prison were taken to mean "those who were then in the prison of sin," or "those who are now in the prison of Hades, but were then alive,"[26] thus avoiding the idea that Christ had literally preached to spirits in the spirit world.

Peter spoke of the Lord's visit to the spirits *in prison* (1 Peter 3:19). Based on modern scripture, *prison* is a comprehensive term that includes the righteous as well as the wicked: "the dead had looked upon the long absence of their spirits from their bodies as a bondage" (D&C 138:50). Even the righteous spirits viewed their existence in the spirit world as living in prison because of their separation from their bodies. Therefore, we may take the term *prison* to refer to the entire spirit world.

"Which sometime were disobedient, when once the longsuffering of God waited in the days of Noah, while the ark was a preparing, wherein few, that is, eight souls were saved by water" (1 Peter 3:20).

Why did Peter single out Noah's generation? One reason may be suggested in "Sanhedrin" (10.3) in the Mishna: "The generation of the Flood have no share in the world to come, nor shall they stand in the judgement." This group was considered by the rabbis to epitomize the most wicked generation in the history of mankind. Therefore, Peter may have been using them as typical of the most wicked, and may have been saying in effect that the Savior's mission to the spirit world embraced all spirits, even the most wicked. The Joseph Smith Translation adds three clarifying words to verse 20: *"Some of whom were disobedient in the days of Noah"* (emphasis added), confirming the sense that Noah's generation was only part—perhaps the most wicked part—of the spirits who benefited from the Savior's mission. A few verses later, 1 Peter 4:6 implies that the gospel was preached to all the dead.

Doctrine and Covenants 138:30–33 records: "[The gospel was preached] even to all the spirits of men; . . . even unto all who would repent of their sins and receive the gospel[;] . . . to those who had died in their sins, without a knowledge of the truth, or in transgression, having rejected the prophets. These were taught faith in God, repentance from sin, vicarious baptism for the remission of sins, the gift of the Holy Ghost by the laying on of hands."

What kind of redemption could those spirits hope for? The foundation of redemption is resurrection, which, by virtue of the Savior's work, all who had had earthly bodies could anticipate. In addition, each could expect to be redeemed "through obedience to the ordinances of the house of God" and receive a reward according to his works (D&C 138:58–59). In Moses 7:38–39 the Lord spoke to Enoch about Noah's generation: "But behold, these which thine eyes are upon shall perish in the floods; and behold, I will shut them up; a prison have I prepared for them. And That [Christ] which I have chosen hath pled before my face. Wherefore, he suffereth for their sins; inasmuch as they will repent in the day that my Chosen shall return unto me, and until that day they shall be in torment" (see also Moses 7:57).

Doctrine and Covenants 138:59 teaches the fate of these spirits: "And after they have paid the penalty of their transgressions, and are washed clean, [they] shall receive a reward according to their works, for they are heirs of salvation." That is, Noah's generation, as well as all those who enter the spirit world unrepentant, experience a cleansing process in the spirit world in preparation for the resurrection. Many of these spirits will receive a degree of redemption in the terrestrial or telestial kingdoms (D&C 76:72–78; 88:99).

"The like figure whereunto even baptism doth also now save us (not the putting away of the filth of the flesh, but the answer of a good conscience toward God,) by the resurrection of Jesus Christ" (1 Peter 3:21).

I include here 1 Peter 3:21, which refers to the Flood being like baptism because it indicates that Peter's attention turned from the subject of those who received the preaching to the subject of baptism: "Which (water) saves you now as a type, namely [of] baptism, which is not the removal of uncleanness of [from] the flesh, but a *covenant* before God of a right mind, through the resurrection of Jesus Christ" (my translation and emphasis).[27] In some areas of the early church,[28] candidates for baptism put off their clothes, entering the baptismal water naked; then, upon emerging from the water, they were clothed with a white garment. For this reason Reicke perceived a possible allusion to a baptismal service in the language of this verse about putting off the uncleanness of the flesh.[29]

Several scholars have observed that 1 Peter, which identifies itself as an epistle, has elements reminiscent of an actual baptismal service. First Peter 3:18–22 contains what looks like a basic creed embracing in a few words a summary of the Lord's ministry: He suffered for sin, died, went in the spirit and preached to spirits in prison (Peter inserted the figure of baptism here), and went to heaven at the right hand of God, having angels, etc., subject to him. One significance of this credal section is that Peter has put baptism in close proximity to the Savior's descent, linking the two. There is a parallel text in 1 Timothy 3:16 that says: "God was manifest in the flesh, justified in the Spirit, *seen of angels* [can refer to spirits], preached unto the Gentiles, believed on in the world, received up into glory" (emphasis added). Reicke comments, "The whole hymn in 1 Timothy 3:16 has . . . great similarity to 1 Peter 3:18–22. From this it is fairly clear that the appearance to the Angel world is a *motif* organically embodied in the Salvation drama."[30]

After study of pre-Nicene paschal and baptismal texts, F. L. Cross concluded that 1 Peter 1:3–4:11 has a baptismal setting and that the rite is understood to have taken place after Peter's words in 1:21, which are: "Who by him do believe in God, that raised him up from the dead, and gave him glory; that your faith and hope might be in God." At this point the person was baptized. Here are the verses that follow the alleged baptism: "Seeing ye have purified your souls in obeying the truth through the Spirit unto unfeigned love of the brethren, see that ye love one another with a pure heart fervently: being born again, not of corruptible seed, but of incorruptible."[31] However, whatever the original setting of the elements in 1 Peter, the significant point is, again, that baptism is seen by many scholars to be the main theme of Peter's letter, and thus the context for the Descent and the preaching to the spirits.

Early Christians associated the Lord's passion and the Descent with baptism, Easter Eve being a popular time to receive baptism. Tertullian (A.D. 160–220) observed, "The Pascha offers the most solemn occasion for Baptism."[32] Of course baptism by immersion is itself a figure of death, being a descent into a watery tomb preceding the deliverance from the amniotic water of spiritual rebirth. But again Peter's association of baptism with the Descent and the preaching to the spirits should be noted because the linking of these two ideas may constitute a cryptic reference to the offering of baptism to the dead, and even to vicarious work for the dead.

"For this cause was the gospel preached also to them that are dead, that they might be judged according to men in the flesh, but live according to God in the spirit" (1 Peter 4:6).

Again scholars debate what "them that are dead" (4:6) means, some wishing to interpret the phrase as pertaining to the spiritually dead. Augustine, Cyril, Bede, Erasmus, Luther, and others took "the dead" to mean "those who were dead in trespasses and sins," i.e., the spiritually dead, or more especially the Gentiles,[33] since these fathers could not assent to the wicked in hell receiving the gospel. But Clement of Alexandria (A.D. 150–215) wrote, "If then He preached [the gospel to those in the flesh that they might not be condemned] unjustly, how is it conceivable that He did not for the same cause preach the gospel to those who had departed this life before His advent?"[34]

EVIDENCES IN JUDAISM AND EARLY CHRISTIANITY OF THE SAVIOR'S REDEMPTIVE WORK IN THE DESCENT

Foreshadowings of the redemptive nature of the Descent appear in the Old Testament (e.g., Zechariah 9:11: "By the blood of the covenant I have sent forth thy prisoners out of the pit wherein is no water." See also Hosea 13:14). In Jewish literature we read not only of the Descent but also of the Messiah's redemptive work in the spirit world. Several apparently Jewish texts, some based on Old Testament passages, contain descriptions of the Lord's work among the spirits of the dead. Justin[35] and Irenaeus[36] quote a passage which they claim was formerly found in the text of Jeremiah (once Irenaeus attributes it to Isaiah) but which they say had been excised by Jewish controversialists. This passage is called the Jeremiah Logion: "The Lord remembered His dead people of Israel who lay in their graves, and went down to preach to them His own salvation." One scholar observes:

"It is strange that no trace of this text is found in the LXX of Jeremias, if what Justin alleges is true, but it must now be admitted (in the light of the Qumran scrolls) that some tampering with controversial texts was practised by the Jews, for the Isaias scroll at 53:11 has a reading which favours the Christian argument (and which is found in the LXX [Septuagint]), but this reading has disappeared from all later Hebrew mss. Moreover, the Greek Fragments of the OT found at Qumran present a type of text which is often in agreement with Justin."[37]

From rabbinic literature, which contained ideas likely contemporary with early Christianity, comes an account of the Lord's visit to the spirit world. The first of two passages from the *Bereshith Rabba* says of the Messiah's appearance at the gates of Gehinnom:

"But when they that are bound, that are in Gehinnom, saw the light of the Messiah, they rejoiced to receive him, saying, He will lead us forth from this darkness, as it is said, 'I will redeem them from Hell, from death I will set them free' (Hosea 13:14); and so says Isaiah (35:10), 'the ransomed of the Lord will return and come to Zion.' By 'Zion' is to be understood Paradise."[38]

And in another passage, "This is that which stands written, 'We shall rejoice and exult in Thee. When? When the captives climb up out of hell, with the Shechinah at their head.'"[39]

An early Christian hymn, dated about A.D. 100, contains

additional insight into the early understanding of the Descent and the redemption of the dead:

"Sheol saw me and was shattered, and Death ejected me and many with me. . . . And I made a congregation of living among his dead . . . and those who had died ran towards Me and cried: 'Son of God, have pity on us . . . and bring us out from the bonds of darkness, and open to us the door by which we shall come out to Thee. . . . Thou art our Savior.' Then I heard their voice. . . . And I placed my name upon their head, because they are free and they are mine."[40]

A second passage from this hymn again represents Christ as speaking:

"And I opened the doors which were closed . . . and nothing appeared closed to me, because I was the opening of everything. And I went towards all my bondsmen in order to loose them; that I might not leave anyone bound or binding. And I gave my knowledge generously, and my resurrection through my love . . . and transformed them through myself. Then they received my blessing and lived, and they were gathered to me and were saved; because they became my members and I was their head."[41]

The combination of Christ's placing his name on the heads of the dead, references to the Savior's work of liberation, and the spirits' reception of Christ's blessing strongly suggest the giving of baptism to the spirits.

In Doctrine and Covenants 138 President Joseph F. Smith recorded that the righteous were assembled in one place waiting for the appearance of the Lord (vv. 11–19, 38–49). This division of spirits is supported by the apocryphal Book of Enoch (dated by Charlesworth as second century B.C. to first century A.D.), which, though not Christian, contains a view of the spirit world current among the Jews of Jesus' time that apparently influenced such New Testament books as Jude and 2 Peter (chapter 2). Enoch, on a tour of the spirit world, asks the attending angel what the hollow places in the rock are. The angel answers, "These beautiful [or 'hollow'—the Greek words are similar] corners [are here] in order that the spirits of the souls of the dead should assemble into them. . . . They prepared these places in order to put them there until the day of their judgment. 'For what reason is one separated from the other?' And he replied and said to me, 'These three have been made in order that the spirits of the dead might be separated. And in the manner in which the souls of the righteous are separated by this spring of water with light upon it, in like manner, the sinners are set apart when they die'"[42]

President Smith wrote, "I beheld that they were filled with joy and gladness, and were rejoicing together because the day of their deliverance was at hand" (D&C 138:15). The joy of the spirits is also supported by Enoch: "Then there came to them a great joy. And they blessed, glorified, and extolled [the Lord] on account of the fact that the name of that [Son of] Man was revealed to them."[43]

A Jewish text dating about A.D. 100 describes the state of the spirits after death: "Did not the souls of the righteous in their chambers ask about these matters, saying, 'How long are we to remain here?'" "[The archangel said] in Hades the chambers of the souls are like the womb. For just as a woman who is in travail makes haste to escape the pangs of birth, so also do these places hasten to give back those things that were committed to them from the beginning."[44]

The Gospel of Nicodemus, a Christian document dating from the time of Justin Martyr, who shows familiarity with it, describes the Savior's advent in Hades:

"We, then, were in Hades with all who have died since the beginning of the world. And at the hour of midnight there rose upon the darkness there something like the light of the sun and shone, and light fell upon us all, and we saw one another. And immediately our father Abraham, along with the patriarchs and the prophets, was filled with joy, and they said to one another: This shining comes from a great light. The prophet Isaiah, who was present there, said: This shining comes from the Father and the Son and the Holy Spirit. This I prophesied when I was still living: . . . the people that sit in darkness saw a great light."[45]

THE ORGANIZATION OF THE RIGHTEOUS TO TAKE THE GOSPEL TO THE WICKED

Doctrine and Covenants 138 maintains that the Savior did not descend personally to the wicked but organized the righteous and gave them authority to engage in the preaching of the gospel to all the dead (vv. 29–30). Later Muhammadan theology also contains a trace of this doctrine: "The righteous ones, who have safely passed the bridge which crosses Hell to Paradise, intercede for their brethren detained upon it. They are sent to Hell to see if any there have faith, and to bring them. These are washed in the Water of Life and admitted to Paradise."[46] In *Yalkut Shim'oni* (Jewish), the godless are rescued from hell by the righteous dead and pass to eternal life, while in the *Zohar,* the righteous or the patriarchs are said to descend to hell to rescue sinners from the place of torment.[47]

The early Christian author Ignatius (A.D. 35–107) writes about the Savior's visit to the prophets, "If these things be so, how then shall we be able to live without him of whom even the prophets were disciples in the Spirit and to whom they looked forward as their teacher? And for this reason he whom they waited for in righteousness, when he came, raised them from the dead."[48]

Another early Christian document refers to the Lord's visit to the righteous: "I have descended and have spoken with Abraham and Isaac and Jacob, to your fathers the prophets, and have brought to them news that they may come from the rest which is below into heaven, and have given them the right hand of the baptism of life and forgiveness and pardon for all wickedness, as to you, so from now on also to those who believe in me. But whoever believes in me and does not do my commandment receives, although he believes in my name, no benefit from it. He has run a course in vain. . . . O Lord, in every respect you have made us rejoice and have given us rest; for in faithfulness and truthfulness you have preached to our fathers and to the prophets, and even so to us and to every man. And he said to us, 'Truly I say to you, you and all who believe and also they who yet will believe in him who sent me I will cause to rise up into heaven, to the place which the Father has prepared for the elect and most elect, (the Father) who will give the rest that he has promised, and eternal life.'"[49]

Rising "up into heaven" reminds us of a striking parallel in Doctrine and Covenants 138:51: "These the Lord taught, and gave them power to come forth, after his resurrection from the dead, to enter into his Father's kingdom, there to be crowned with immortality and eternal life."

Other early religious literature adds some interesting details to the idea of the righteous taking the gospel to the dead. Hippolytus writes, "John the Baptist died first, being dispatched by Herod, that he might prepare those in Hades for the gospel; he became the forerunner there, announcing even as he did on this earth, that the Savior was about to come to ransom the spirits of the saints from the hand of death."[50]

Again, in a sixth-century manuscript, Hippolytus distinguishes between those who saw the Savior and those who only heard his voice, the voice being figurative perhaps for hearing the gospel from his authorized servants:

"Oh, Thou only-begotten Son among only-begotten sons, and All in all! Seeing that the multitude of Holy Souls was deep down and had been deprived of a Divine visit long enough, the Holy Spirit had previously said that they should be the object of a meeting with the Divine Soul, saying: 'His form we have not seen, but His voice we have heard.' For it behoved Him to go and preach also to those who were in Hell, namely those who had once been disobedient."[51]

One more example will illustrate the righteous spirits taking the gospel to the wicked spirits and will also provide a transition to the topic of vicarious work for the dead. The *Shepherd of Hermas* (first century) was, according to the fourth-century church historian Eusebius, considered by some valuable for instruction in the church, and was quoted by some of the most ancient writers.[52] Hermas saw in a vision that the apostles took the gospel into the spirit world so that the dead might receive the seal of baptism:

"These apostles and teachers, who preached the name of the Son of God, having fallen asleep in the power and faith of the Son of God, preached also to those who had fallen asleep before them, and themselves gave to them the seal of the preaching [baptism]. They went down therefore with them into the water and came up again, but the latter went down alive and came up alive, while the former who had fallen asleep before, went down dead but came up alive."[53]

Clement of Alexandria also cited this passage, commenting "that it was necessary for the apostles to be imitators of their Master on the other side as well as here, that they might convert the gentile dead as he did the Hebrew."[54] In another place, citing Hermas again, he wrote that "Christ visited, preached to, and baptized the just men of old, both gentiles and Jews, not only those who lived before the coming of the Lord, but also those who were before the coming of the Law . . . such as Abel, Noah, or any such righteous man."[55]

Clement's observation recalls Doctrine and Covenants 138:40–41 and the description of all the righteous prophets, including Abel and Noah, who waited in that assembly for the Savior's advent.

VICARIOUS WORK FOR THE DEAD

The writer of Hermas, quoted in the passage above, obviously treated two kinds of death, spiritual and physical, but his poetic writing is not clear, and one might easily find in this passage only a description of apostles giving baptism in the spirit world. But note the sentence, "The latter [the baptizers] *went down alive and came up alive,* while the former [those baptized] . . . went down dead but came up alive." The wording suggests that the former group—the baptizers—is physically alive and the latter—those baptized—physically dead; the sentence therefore cryptically alludes to the practice of vicarious baptism for the dead.

Why is baptism for the dead only hinted at in early writings? It was likely a restricted part of the Savior's teaching. The New Testament contains many references to mysteries of the kingdom that were shared only in the Savior's most intimate circle.[56] Some of the early Fathers exhibit this same sense of secrecy about the special doctrines that the Savior taught, probably during the mysterious forty-day ministry (Acts 1:3).

With regard to secrecy, Clement of Alexandria says of himself that he writes "in a studied disorder"[57] and has "here and there interspersed the dogmas which are the germs of true knowledge, so that the discovery of the sacred traditions may not be easy to any one of the uninitiated."[58] As a result of the secrecy, after the first century and the demise of the authority to administer baptism, the doctrine of vicarious baptism is hardly referred to, and then only with confusion, since those who knew the truth were probably not writing it down in any detail. Finally no one could remember just what it was all about. However, traces linger in the literature. Epiphanius, Bishop of Salamis (A.D. 347–403), wrote:

"From Asia and Gaul has reached us the account [tradition] of a certain practice, namely, that when any die without baptism among them, they baptize others in their place and in their name so that, rising in the resurrection, they will not have to pay the penalty of having failed to receive baptism, but rather will become subject to the authority of the Creator of the World. For this reason this tradition which has reached us is said to be the very thing to which the Apostle himself refers when he says, 'If the dead rise not at all, what shall they do who are baptized for the dead?' Others interpret the saying finely, claiming that those who are on the point of death, if they are catechumens, are to be considered worthy, in view of the expectation of baptism which they had before their death. They point out that he who has died shall also rise again, and hence will stand in need of that forgiveness of sins that comes through baptism."[59]

Later writers thought that only the heretics had practiced baptism for the dead. Many heretics had. For example, the Marcionites would lay a catechumen (candidate for baptism) who had just died upon a bed and lay a living person under his bed. Then they would ask the corpse if he wished to receive baptism, and the living person would reply that he did wish to. Then the living person would be baptized for the dead one.[60]

It was a short step from baptism for the dead to baptism *of* the dead. Greek Canon 20 from the Council of Carthage in A.D. 419 (reporting a Council in A.D. 379) reads, "It also seemed good that the Eucharist should not be given to the bodies of the dead. For it is written, 'Take, Eat,' but the bodies of the dead can neither 'take' nor 'eat.' Nor let ignorance of the presbyters baptize those who are dead."[61] Some of our Latter-day Saint literature has claimed that a Council of Carthage banned baptism for the dead, but in fact it was the baptism of corpses that was banned.

The idea that the living might do something efficacious for the dead was not new in Israel. In 2 Maccabees 12:42 we read, "The noble Judas called on the people to keep themselves free from sin, for they had seen with their own eyes what had happened to the fallen because of their sin. He levied a contribution from each man, and sent the total of two thousand silver drachmas to Jerusalem for a sin-offering—a fit and proper act in which he took due account of the resurrection. For if he had not been expecting the fallen to rise again, it would have been foolish and superfluous to pray for the dead. But since he had in view the wonderful reward reserved for those who die a godly death, his purpose was a holy and pious one. And this was why he offered an atoning sacrifice to free the dead from their sin."

Montefiore quotes from fifth-century rabbinic literature on the redemption of man from hell:

"Hence we learn that the living can redeem the dead. Hence we have established the rite of holding a memorial service for the dead on the Day of Atonement. . . . For God brings them out of Sheol and they are shot forth as an arrow from a bow. Straightway a man becomes tender and innocent as a kid. God purifies him as at the hour of his birth, sprinkling pure water on him from a bucket. Then man grows up and increases in happiness like a fish which draws happiness from the water. So is a man baptized every hour in rivers of balsam, milk, oil, honey: he eats of the tree of life continuously, which is planted in the division [*Mehizah,* a term referring to the divisions in Paradise] of the righteous and his body reclines at the (banquet) table of every single saint and he lives for eternity."[62]

We must not pass by the Lord's dialogue with Peter where he promises the chief apostle that he will give him keys so that the gates of hell cannot prevail against the Church (Matthew 16:18–19). The image means that the gates of hell would not be strong enough to prevent the Savior's entrance and execution of his redeeming work among the dead, nor would they be strong enough to retain the believing dead for whom proxy work would be done on earth. By the light of the Restoration, this passage is a clear reference to the Lord's intent to empower the Church to perform sealing ordinances for the living and the dead that would transcend death and extend into the eternities.

Here we have seen that abundant writings from the early Christian period, especially from the earliest Church writers, support the doctrine that the Son of God journeyed to the spirit world and there performed a work of enormous magnitude which, with his mortal ministry, offered redemption to the entire family of God. At the same time, it is apparent that the forces of apostasy clouded this precious doctrine in confusion. One characteristic of apostasy is that it seeks to narrow understanding of the scope of the Lord's redemptive work, his grace, his love, and his power. But the truth about the heart of God is reflected in the prophet Enoch's reaction to the plight of the miserable wicked:

"Enoch . . . looked upon their wickedness, and their misery, and wept and stretched forth his arms, and his heart swelled wide as eternity; and his bowels yearned; and all eternity shook. . . . And as Enoch saw this [the swallowing up of the wicked in the floods at the time of Noah], he had bitterness of soul, and wept over his brethren, and said unto the heavens: I will refuse to be comforted" (Moses 7:41, 44).

But the Lord responded: "Lift up your heart, and be glad. . . . I am Messiah, the King of Zion, the Rock of Heaven, which is broad as eternity; whoso cometh in at the gate and climbeth up by me shall

never fall. . . . Righteousness will I send down out of heaven; and
truth will I send forth out of the earth. . . . And righteousness and
truth will I cause to sweep the earth as with a flood" (Moses 7:44, 53,
62).

Doctrine and Covenants 138 is a fulfillment of that promise of
truth from heaven. It is not new doctrine, but a restoration of the
knowledge that God had given the ancients.

Notes

Revised from *The New Testament and the Latter-day Saints*, Sperry Symposium 1987
(Orem, Utah: Randall Book, 1987), 295–317.

1. One particularly good translation of Dante Alighieri's *The Inferno* is John Ciardi's (New York: New American Library, 1954), 4:15–60, pages 50–51.
2. J. A. MacCulloch, *The Harrowing of Hell* (Edinburgh: T&T Clark, 1930), 83; MacCulloch observes that the earliest patristic (early Church "fathers") references to the Descent occur in the Epistles of Ignatius and are made in such a way as to show that he is treating a well-known belief. See also Magnes c. 9; Phila c. 5, c 9; Trall c. 9.
3. Polycarp, in his Epistle to the Philippians (1.2), implies the Descent by his citation of Peter's sermon in Acts 2:24 (MacCulloch, *The Harrowing of Hell*, 84).
4. Dialogue with Trypho, c. 72.
5. Adv. Haer. 4.27.2; Irenaeus says he heard this doctrine that Jesus "descended into the regions beneath the earth . . . preaching the remission of sins received by those who believe in Him" from "a certain presbyter, who had heard it from those who had seen the apostles, and from those who had been their disciples" (4.27.1–2).
6. De Resur. Carnis, 43, 44; de Anima 7, 55, 58; adv. Marcionem, 4.34.
7. Edward G. Selwyn, *The First Epistle of St. Peter* (London: Macmillan, 1958), 322n.
8. Brigham Young taught that the spirit world is here on the earth: "When you lay down this tabernacle, where are you going? Into the spiritual world. Are you going into Abraham's bosom? No, not anywhere nigh there but into the spirit world. Where is the spirit world? It is right here." See Brigham Young, *Discourses of Brigham Young*, sel. John A. Widtsoe (Salt Lake City: Deseret Book Co., 1971), 376.
9. *Encyclopedia of Religion and Ethics*, 13 vols., ed. J. H. Hastings (New York: Scribners, 1951), 4:654–63.
10. Richard L. Anderson, *Understanding Paul* (Salt Lake City: Deseret Book Co., 1983), 413.
11. Charles Bigg, *Critical and Exegetical Commentary on the Epistles of St. Peter and St. Jude* (Edinburgh: T&T Clark, 1902). William J. Dalton, *Christ's Proclamation to the Spirits: A Study of 1 Peter 3:18–4:6* (Rome: Pontifical Biblical Institute, 1965). MacCulloch, *The Harrowing of Hell*. Hugh Nibley, "Baptism for the Dead in Ancient Times," reprinted from *Improvement Era*, 19 December 1948–19 April 1949 (Provo: Foundation for Ancient Research and Mormon Studies, 1993. Bo Reicke, *The Disobedient Spirits and Christian Baptism* (New York: AMS Press, 1946). Selwyn, *First Epistle of St. Peter*.
12. Bigg, *Critical and Exegetical Commentary*, 162.

13. Augustine, de Haer. 79; St. Gregory, Ep. 15; J. N. D. Kelly, *Early Christian Creeds* (New York: David McKay Co., 1972), 381.
14. Reicke, *Disobedient Spirits and Christian Baptism,* 14.
15. Kelly, *Early Christian Creeds,* 379.
16. Ibid.
17. Ibid., 383.
18. Ibid., 378.
19. Reicke, *Disobedient Spirits and Christian Baptism,* 14.
20. Ibid., 16.
21. Samuel Fuller, *Defense of the Version of King James I,* "The Spirits in Prison" (New York: T. Whittaker, 1885).
22. *Words of Joseph Smith,* 370; emphasis added.
23. C. Cels. 2.43.
24. De Antichr. c 26.
25. De Haeresibus, 79.
26. Ad Evodium, Ep. 164.
27. Compare Mosiah 3:19, "natural" and "spiritual."
28. Probably the earliest reference to nudity in baptism is found in Hippolytus, Apostolic Tradition 21.3 (early 3rd century) and possibly in the Syria Didascalia (c. 250) where nudity in baptism is strongly implied. The practice of nudity in baptism may stem from the period when baptism became confused with other sacred ordinances. Several phrases from Paul were later interpreted to allude to the practice of nudity: Galatians 3:27, Colossians 3:9–10, Ephesians 4:22–24, etc.
29. Reicke, *Disobedient Spirits and Christian Baptism,* 189–90.
30. Ibid., 234; compare the baptismal covenant in Mosiah 18:10.
31. F. L. Cross, *1 Peter: A Paschal Liturgy* (London, 1954).
32. De Baptismo, 19.
33. Bigg, *Critical and Exegetical Commentary,* 171.
34. Stromata 6.6.48.
35. Dial. 72.
36. In five places: adv. Haer. 3. 22.1; 4.3.61; 4.50.1; 4.55.3; 5.31.1H.
37. J. H. Crehan, *A Catholic Dictionary of Theology* (London; New York: Nelson, 1967), 163.
38. Most easily accessible in MacCulloch, *The Harrowing of Hell,* 31; see also *Encyclopedia of Religion and Ethics,* 4:653.
39. Ibid.
40. *Odes of Solomon* 42:11–20 in J. H. Charlesworth, *The Odes of Solomon, The Syriac Texts* (Missoula, MT: Scholars Press, 1977).
41. *Odes of Solomon* 17:9–16.
42. Book of Enoch 22:8–10.
43. Book of Enoch 68:27.
44. 4 Ezra 4:35, 42; compare *Odes of Solomon* 24:5.
45. Hennecke, Schneemelcher, and Wilson, *New Testament Apocrypha,* 2 vols. (Philadelphia: Westminster Press, 1963) 1:471.
46. This reference is most easily available in MacCulloch, *The Harrowing of Hell,* 32, and *Encyclopedia of Religion and Ethics,* 4:653.
47. MacCulloch, *The Harrowing of Hell,* 31; *Encyclopedia of Religion and Ethics,* 4:653.
48. Magn. 9.2.

49. *Epistula Apostolorum* (c. A.D. 100–150), 27–28, in Hennecke et al., *New Testament Apocrypha,* 209–10.

50. De Antichr. c. 45; see also Origen in two places, In Luc. Homil. c. 4., and In Evang. John 2.30.

51. Reicke, *Disobedient Spirits and Christian Baptism,* 24.

52. *Ecclesiastical History* 3.3.6.

53. Sim. 9.16.5.

54. Stromata 6.6.

55. Ibid. 2.9.

56. Matthew 13:11; Mark 4:11; 2 Corinthians 12:2–4; 1 Corinthians 2:6; See also Eusebius, *Ecclesiastical History* 2.1.4 and Nibley, *Baptism for the Dead in Ancient Times,* part 1, 1–5.

57. Stromata 6.2.1.

58. Stromata 7.110.4.

59. Adv. Haer. 1.28.6.

60. John Chrysostome, Homil. 40 in 1 Cor.

61. "The Seven Ecumenical Councils," in Philip Schaff and Henry Wace, eds., *Nicene and Post-Nicene Fathers,* 14 vols. (Grand Rapids, Mich.: Eerdmans Publishing Co., 1989, 1994) 14:451.

62. Tanhuma, "Ha'asinu," I, f. 339b. C. G. Montefiore, *A Rabbinic Anthology* (New York: Schocken Books, 1974), 675.

HEBREWS: TO ASCEND THE HOLY MOUNT

Hebrews is, to use Paul's¹ words, "strong meat" (Hebrews 5:14). Paul wants to preach strong meat, but he addresses members who are reluctant to digest it (v. 12). Nevertheless, he broaches doctrines that deal with the upper reaches of spiritual experience and Melchizedek Priesthood temple ordinances. My purpose will be to identify several passages that have relevance to temple ordinances. Paul's letter might be divided into two main ideas: the *promise* of the temple and the *price* exacted to obtain the promise. At several points I will add the Prophet Joseph Smith's commentary, without which much of the temple significance of the apostle's remarks in Hebrews would elude us.

THE PROMISE

Paul urges the Hebrews, "Let us go on unto *perfection;* not laying *again* the foundation of repentance . . . and of faith" (Hebrews 6:1).² They had tarried too long in the foothills of spiritual experience. Having "tasted of the heavenly gift, . . . the good word of God, and the powers of the world to come" (vv. 4–5), they could no longer delay resuming the climb lest they lose the *promise.* Paul warns, "Be not slothful, but followers of them who through faith and patience inherit [or, *are* inheriting] the promises" (v. 12).

The *promise* that Paul refers to repeatedly is that same promise explained in Doctrine and Covenants 88:68–69: "Therefore, sanctify yourselves that your minds become single to God, and the days will come that you shall see him; for he will unveil his face unto you, and it shall be in his own time, and in his own way, and according to his own will. Remember the great and last *promise* which I have made unto you." Paul uses several different terms in Hebrews for the experiences associated with this promise:[3] for example, *obtaining a good report* (11:39), *entering into the Lord's rest* (4:3, 10), *going on to perfection* (6:1), *entering into the holiest* (10:19), *being made a high priest forever* (7:17), *knowing the Lord* (8:11; D&C 84:98), *pleasing God* (Hebrews 11:5), *obtaining a witness of being righteous* (11:4), and *having the law written in the heart* (8:10; 10:16; Jeremiah 31:31–34).[4] He speaks of boldly pursuing the fulfillment of the promise: Grasp, he says, the hope that is set before you, which enters behind the veil, where Jesus, as a forerunner, has already entered (Hebrews 6:18–20).

Paul compares these Israelites to their ancestors of twelve hundred years earlier. He refers to the early Israelites' rejection of God's invitation to enter into his rest as the "provocation"; that is, Israel provoked God by refusing to enter his presence. Paul quotes from Psalm 95:8–11: "Harden not your hearts, as in the *provocation,* in the day of temptation in the wilderness: when your fathers tempted me, proved me, and saw my works forty years. Wherefore I was grieved with that generation, and said, . . . they have not known my ways. So I sware in my wrath, They shall not enter into my rest" (Hebrews 3:8–11).

In this Exodus account to which Paul alludes, the children of Israel gazed at the quaking, smoking, fiery mount and refused to exercise the faith to go up. The upper reaches of the mount are, to be sure, not for the fainthearted. The frightened Israelites foolishly told Moses *to go on their behalf* (Exodus 20:18–21). The Lord, referring to the Melchizedek Priesthood as the key to God's presence, explains in modern revelation what it was that Israel rejected: "For without this [priesthood] no man can see the face of God, even the Father, and live. Now this Moses plainly taught to the children of Israel in the wilderness, and sought diligently to sanctify his people that they might behold the face of God; but they hardened their hearts and could not endure his presence; therefore, the Lord . . . swore that they should not *enter into his rest* while in the wilderness, which rest is the fulness of his glory. Therefore, he took Moses out of their midst, and the Holy Priesthood also" (D&C 84:22–25).

It is sobering that it was unnecessary for the Israelites to wander in the wilderness for forty years. Had they exercised faith in Jehovah, who is mighty to deliver, they might have abbreviated those trials and entered speedily into the promised land and into a Zion, even a translated society like Enoch's or Melchizedek's (D&C 105:2–6). But, Paul laments, the early Israelites refused to enter because of *unbelief* (Hebrews 3:19). He says, "Let us therefore fear, lest, a promise . . . of entering into his rest, any of you should . . . come short of it. . . . For we which have believed *do* enter into rest" (Hebrews 4:1, 3). Among Paul's fellows were those who were even then entering into the Lord's rest.

The Joseph Smith Translation of Exodus 34 increases our vocabulary for what it was that Israel rejected: "I will take away the priesthood out of their midst; therefore my *holy order,* and *the ordinances* thereof, shall not go before them; for *my presence* shall not go up in their midst" (v. 1). The Prophet Joseph remarked on Israel's rejection, using yet another term for the loss, that is, the term *last law:*

"God cursed the children of Israel because they would not receive the *last law* from Moses. . . . When God offers a blessing or knowledge to a man, and he refuses to receive it, he will be damned. The Israelites prayed that God would speak to Moses and not to them; in consequence of which he cursed them with a carnal law. . . . [But] the law revealed to Moses in Horeb never was revealed to the children of Israel as a nation."[5]

When God gives the Saints the Melchizedek Priesthood, which is the power and authority to ascend into the presence of God through temple ordinances, they must come or be damned.

The Aaronic Priesthood retained the keys to the ministry of

angels but not to the presence of God (D&C 84:26). Hebrews opens with a discussion of Christ's superiority over ministering angels. Paul's point is that even though Israel chose a law of intermediaries, that is, the ministering of angels, they must not value angels over the direct presence of God. They had chosen the keys to an anteroom but rejected those to the throne room itself.

The history of Israel is punctuated by their preference for intermediaries over God himself. One scholar notes, "Once the immediacy of early prophecy comes to an end, the angels serve to mediate the secrets of nature, the heavenly world and the last age."[6] Josephus reports that the Essenes had a preoccupation with the secret names of angels,[7] and the fascination of the mystical kabbalistic Jews with angelic hierarchies is well known. The early Christian interposition of saints between God and man is another form of substitution of intermediaries for God himself.

One may indeed receive keys to discern and control angelic visitations (D&C 129). Joseph Smith taught that there were keys of the kingdom, "certain signs and words by which false spirits and personages may be detected from true, which cannot be revealed to the Elders till the Temple is completed. . . . There are signs . . . the Elders must know . . . to be endowed with the power, to finish their work and prevent imposition."[8] But the applicant for exaltation must *exceed* the right to the ministry of angels in order to regain the presence of God. The Lord said to the Church in this dispensation with respect to angels assisting in the redemption of Zion: "Let not your hearts faint, for I say not unto you as I said unto your fathers: Mine *angel* shall go up before you, but not my presence [Exodus 33:2–3]. But I say unto you: Mine angels shall go up before you, and also *my presence*" (D&C 103:19–20).

In attempting to persuade the Hebrew members of the superiority of the Melchizedek law over the Aaronic, Paul implies that an order of holy beings prevails in the eternal worlds that the Saints are called to enter. Christ belongs to this order, as did Melchizedek. Paul deals in three places with Melchizedek: chapters 5, 7, and 11 (although in chapter 11 Paul does not name him). Though man is created a little lower than the angels here on earth, yet his destiny is to put all in subjection under him, as Christ did, who brings "many sons unto glory" (Hebrews 2:10). "Salvation is nothing more or less," said Joseph Smith, "than to triumph over all our enemies and put them under our feet and when we have power to put all enemies under our feet in this world and a knowledge to triumph over all evil spirits in the world to come, then we are saved, as in the case of Jesus."[9] Alma teaches that "many, exceedingly great many," have entered into this holy order, Melchizedek being prototypical of them (Alma 13:12, 18).[10]

Paul maintains that the Levitical law never could have brought its adherents into the Holy of Holies (see, for example, Hebrews 7:11). Under the Levitical law only the high priest entered there, and then only once a year. Therefore, so long as the Levitical or Mosaic law still stood, the way into the sanctuary necessarily remained veiled (Hebrews 9:8). Christ rent the veil to the Holy of Holies to make entrance behind the veil possible, not for just one high priest, but for a whole kingdom of high priests (Hebrews 10:20; Exodus 19:6).

Paul alludes to three levels of priesthood power. The Levitical, which could never make anyone perfect; Abraham's patriarchal power, which embraces eternal marriage; and Melchizedek's, which was a power greater still than Abraham's, "even power of an endless life, of which [order] was our Lord Jesus Christ, which [order] also Abraham [later] obtained by the offering of his son Isaac. [Abraham's] power [was not that] of a prophet nor apostle nor patriarch only, but of king and priest to God, to open the windows of Heaven and pour out the peace and law of endless life to man, and no man can attain to the joint heirship with Jesus Christ without being administered to by one having the same power and authority of Melchizedek"[11] (see also JST, Genesis 14:40; Hebrews 7:6, 17). "If a man gets the fulness of God he has to get [it] in the same way that Jesus Christ obtained it and that was by keeping all the ordinances of the house of the Lord."[12] Thus, through obedience to

Melchizedek Priesthood temple ordinances, fallen man and woman may obtain the order of Melchizedek, Abraham, and Christ.

But Paul perceived that his flock could not digest the full truth about Melchizedek's priesthood power (Hebrews 5:11), so he alluded obliquely to him in Hebrews 11:33–34. That the allusion is to Melchizedek is clear from the Joseph Smith Translation of Genesis 14:26, which describes Melchizedek in nearly identical wording, and then continues by saying that Melchizedek had the priesthood power of translation by which many of the citizens of his city obtained translation. Paul mentioned earlier in this chapter (Hebrews 11:8–10) that Abraham, Isaac, and Jacob also sought an inheritance in this heavenly city of translated beings; that is, they sought to be translated and to join the city of Enoch, as had those who became Saints "during the nearly seven hundred years from the translation of Enoch to the flood of Noah."[13]

THE PRICE

Paul refers repeatedly to suffering and sacrifice. It is at this point that we sense why the saints of any day would tremble at ascending the holy mount. Temple covenants of sacrifice are quite comprehensive. Paul defines *high priest* as one who makes sacrifices for others (Hebrews 5:1), referring to the function of the high priest in the Mosaic temple, but perhaps more broadly to all high priests. After all, the veil that Christ, the great high priest, rent for us was the veil of his own flesh, not only opening the way for us into the holiest, but showing how comprehensive is the sacrifice required to follow him and obtain his order (Hebrews 10:19–20).

We have the ambiguous passage in Hebrews 5:7–9 that seems to refer at the same time both to Christ and Melchizedek: "Though he were a Son, yet learned he obedience by the things which he suffered" (v. 8). Sometimes this passage is misinterpreted to mean that Christ or Melchizedek had to suffer the consequences of not obeying before they learned to obey. Rather, the sense is that they were willing to submit to *suffering anything necessary* in order to come up to the full measure of obedience to God, and by so sacrificing, achieved perfection. Elder Spencer W. Kimball says similarly: "To each person is given a pattern—obedience through suffering, and perfection through obedience."[14]

It is not just any sacrifice or suffering that suffices, but that which is necessary to fulfill what God requires (2 Nephi 31:9; 1 Samuel 15:22—obedience is "better than sacrifice"). Nevertheless, the sufferings and sacrifices of the saints become, as Peter says, more precious than fine gold (1 Peter 1:7; 4:13). John Taylor wrote that Joseph Smith spoke in a similar vein to the twelve apostles: "You will have all kinds of trials to pass through. And it is quite as necessary that you be tried as it was for Abraham and other men of God. . . . God will feel after you, and He will take hold of you and wrench your very heart strings, and if you cannot stand it you will not be fit for an inheritance in the Celestial Kingdom of God"[15] (D&C 97:8).

How can one press forward in the midst of sacrificing and suffering? The Prophet Joseph answers in the *Lectures on Faith:* "They are enabled by faith to lay hold on the promises which are set before them, and wade through all the tribulations and afflictions to which they are subjected by reason of the persecution from those who know not God, and obey not the gospel of our Lord Jesus Christ . . . believing that the mercy of God will be poured out upon them in the midst of their afflictions, and that he will compassionate them in their sufferings, and that the mercy of God will lay hold of them and secure them in the arms of his love."[16]

"Let us here observe, that a religion that does not require the sacrifice of all things never has power sufficient to produce the faith necessary unto life and salvation. . . . It was through this sacrifice [of all earthly things], and this only, that God has ordained that men should enjoy eternal life; and it is through the medium of the sacrifice of all earthly things that men do actually know that they are doing the things that are well pleasing in the sight of God. When a man has offered in sacrifice all that he has for the truth's sake, not even withholding his life, and believing before God that he has been called to make this sacrifice because he seeks to do his will, he does know, most assuredly, that God does and will accept his sacrifice and offering, and that he has not, nor will not seek his face in vain. Under these circumstances, then, he can obtain the faith necessary for him to lay hold on eternal life."[17]

Referring to Paul's well-known statement about our fathers not being able to be perfect without us, nor we without them, I quote the Joseph Smith Translation rewording: "God having provided [Greek *provided beforehand*] some better things for them through their sufferings, *for without sufferings* they could not be made perfect" (JST, Hebrews 11:40). The Prophet Joseph stated this idea in another place: "Men have to suffer that they may come upon Mount Zion [and be] exalted above the heavens."[18]

The Prophet Joseph used this same verse as a proof text for temple work for the dead. Scripture is susceptible of multiple interpretations, and, in this case, the ideas of suffering, of sacrifice, and of sealing are part of the larger picture of sanctification. In fact, the sacrifice that the sons of Levi will offer up is identified with the book of remembrance of the dead in Doctrine and Covenants 128:24, the section in which the prophet speaks of the welding link necessary with ancestors and makes reference to Hebrews 11:40.

This much is clear then: life is not granted to *please* us or to *satisfy* our telestial ideas of what life should be; but rather it is given and arranged to *develop, refine,* and *reveal* to us what remains in the sanctifying process. The measure of ourselves is best taken as we come up against the sacrifice that full consecration requires in the course of life. Spiritual refinement, as well as coming to the greater light and knowledge, are accomplished through an all-encompassing sacrifice, usually made over time, similar in our own limited sphere to the Savior's sacrifice in his greater sphere; that is, the sacrifice must fill the whole sphere, no matter how great or small. As he drank the cup his Father gave him, so the Saints drink what the Lord Jesus gives them. The Savior's cup was not that he be ministered to but to minister and to give his life as a ransom for many (Matthew 20:28).

Still on the theme of suffering, Paul makes an enigmatic remark,

"Others were tortured, not accepting deliverance [from trials and suf-ferings]; that they might obtain a better resurrection" (Hebrews 11:35). The Prophet Joseph defines *deliverance* as *translation* and identifies the place of habitation of those translated as "that of the ter-restrial order and a place prepared for such characters; . . . [these who were translated] he held in reserve to be ministering angels unto many planets, and who as yet have not entered into so great a fulness as those who are resurrected from the dead."[19]

The confusion from Paul's comment lessens with the realization that he is treating translation as a sort of resurrection (in the sense of a regeneration) but not the literal resurrection, that "better resurrec-tion." He makes a distinction between the temporary, lesser state of translation and the greater change of resurrection ("a better resurrec-tion"). In a fragmentary and undeveloped commentary on Paul's statement, the Prophet Joseph explains that some who were worthy to receive deliverance from their trials and suffering by being trans-lated chose rather to prolong the labors of their ministries in the mortal condition, even submitting to tortures, so as to obtain an immediate rest after their earth life and go directly to that greater rest of resurrection instead of tarrying in that intermediate, terrestrial state (translation) and then later receiving their resurrection. Those who made this choice apparently wished to receive the greater refinement of sacrifices in the telestial condition.[20]

At the end of Hebrews, Paul returns to the mighty promises asso-ciated with the ascent of the holy mount: he says the mount that Israel in his day confronts is not physical or earthly like the one their fathers refused to ascend; rather, the saints' privileges are to "come unto mount Sion, and unto the city of the living God, the heavenly Jerusalem, and to an innumerable company of angels, to the general assembly and church of the firstborn, which are written in heaven, and to God the Judge of all, and to the spirits of just men made per-fect, and to Jesus" (Hebrews 12:22–24). Then soberly he adds, "*See that ye refuse not him that speaketh*" (v. 25). Joseph Smith said in further commentary on this passage:

"The Hebrew church 'came unto the spirits of just men made

perfect, . . . unto . . . angels, unto God . . . , and to Jesus Christ . . . ;' but what they learned, has not been, and could not have been written. What object was gained by this communication with the spirits of the just, etc.? It was the established order of the kingdom of God—the keys of power and knowledge were with them [the angels] to communicate to the Saints— . . . What did they learn by coming to the spirits of just men made perfect? Is it written? No! [It can't be written.] The spirits of just men are made ministering servants to those who are sealed unto life eternal and it is through them that the sealing power comes down."[21]

The urge to know the mysteries of godliness is not necessarily an idle curiosity; rather, it is a divine drive to acquire that level of godly power modeled by Christ and others of his holy order. It is in addition the means of increasing one's power to bring others to Christ: "And if thou wilt inquire, thou shalt know mysteries which are great and marvelous; . . . that thou mayest bring many to the knowledge of the truth" (D&C 6:11; see also Alma 26:22).

The insight lying interlinearly in Hebrews and in the Prophet Joseph's remarks suggests that men and women may do what Christ did by learning and applying eternal law, entering by conscious knowledge and power into their exaltation. This life, Paul seems to say, as does Amulek, is the time for men to prepare to meet God (Alma 34:32). We may have "*boldness* to enter into the holiest by the blood of Jesus" (Hebrews 10:19). This achievement requires a faith that seems to border on audacity. But he reassures his readers that, as the Savior is so abundantly able to succor his people, we may "therefore come *boldly* unto the throne of grace, that we may obtain mercy, and find grace to help in time of need" (Hebrews 4:16).

The Prophet Joseph wrote an impassioned letter to his uncle about these stirring possibilities, quoting Hebrews 6: "[Paul said,] 'We have . . . an anchor of the soul, both sure and steadfast and which entereth into that within the veil' [Hebrews 6:18–19]. Yet [Paul] was careful to press upon them the necessity of continuing on until they . . . might have the assurance of their salvation confirmed to them by an oath from the mouth of him who could not lie. For that seemed to be the example anciently, and Paul holds it out to his brethren as an object attainable in his day. And why not? . . .

"If the Saints in the days of the apostles were privileged to take the [earlier] Saints for example and lay hold of the same promises . . . [that is] that they were sealed there . . . will not the same faithfulness, the same purity of heart, and the same faith bring the same assurance of eternal life—and that in the same manner—to the children of men now in this age of the world? . . .

" . . . And have I not an equal privilege with the ancient saints? And will not the Lord hear my prayers, and listen to my cries, as soon as he ever did to theirs if I come to him in the manner they did?"[22]

Many Saints in the Church hunger and thirst after greater righteousness and spiritual experience, just as our father Abraham did (Abraham 1:2). The hunger is our birthright. Nevertheless, it is common to discourage such people out of fear that they will go off the track somehow in their pursuit, and of course that danger continually presents itself. Old Scratch, as one of my friends calls the adversary, is always lurking behind a tree.

But the opposite risk is that members will straggle in the foothills of spiritual experience, as Israel has repeatedly done. So Paul says, "[Exhort] one another: and so much the more, as ye see the day approaching" (Hebrews 10:25); "for ye have need of patience, that, after ye have done the will of God, ye might receive the promise. For yet a little while, and he that shall come will come, and will not tarry" (vv. 36–37). Paul's letter is a powerful call to pay the price, to obtain the promise in spite of earth or hell, and to come all the way up the holy mount to the Lord Jesus Christ.

Notes

Revised from *Temples of the Ancient World*, ed. Donald W. Parry (Salt Lake City: Deseret Book Co., 1994), 479–91.

1. The basic premise in this paper is that the Apostle Paul is the author of Hebrews, a fact that the Prophet Joseph Smith acknowledged on several occasions.

2. The emphasis in this verse and in subsequent scriptural passages in this chapter has been added by the author.

3. I have paraphrased some of these examples.

4. Joseph Smith says that the law written in the heart will be fulfilled when the Saints' callings and elections are made sure and when they receive the Second Comforter (see *Words of Joseph Smith*, 19, n. 9).

5. *Teachings of the Prophet Joseph Smith*, 49, 50; emphasis added.

6. *Theological Dictionary of the New Testament,* ed. Geoffrey W. Bromiley, 10 vols. (Grand Rapids, Mich.: Eerdmans, 1964–76), 1:80–81.

7. Ibid.

8. *Words of Joseph Smith*, 20–21.

9. Ibid., 200.

10. See Robert Millet, "The Holy Order of God (Alma 13)," in *The Book of Mormon: Alma, the Testimony of the Word,* ed. Monte S. Nyman and Charles D. Tate, Jr. (Provo: BYU Religious Studies Center, 1992), 61–88.

11. *Words of Joseph Smith*, 245.

12. Ibid., 213.

13. Bruce R. McConkie, *Doctrinal New Testament Commentary,* 3 vols. (Salt Lake City: Bookcraft, 1973), 3:202.

14. Edward L. Kimball, ed., *Teachings of Spencer W. Kimball* (Salt Lake City: Bookcraft, 1982), 168.

15. As quoted by John Taylor, *Journal of Discourses*, 24:197.

16. *Lectures on Faith*, 4:14–15.

17. Ibid., 6:7.

18. *Words of Joseph Smith*, 245.

19. Ibid., 41–42.

20. Ibid.

21. Ibid., 253, 254 (cf. D&C 77:11).

22. Letter to Silas Smith, 26 September 1833, in *Personal Writings of Joseph Smith,* ed. Dean C. Jessee (Salt Lake City: Deseret Book Co., 1984), 299–301.

Chapter   11

DOCTRINES RESTORED: ATONEMENT, MARRIAGE, AND WOMANHOOD

Among the early Christians, a philosophy arose called *asceticism*[1] that ultimately dismantled several key doctrines of the true Church. Asceticism holds that the material world is devoid of spiritual value and that only by renouncing this world and, in particular, rejecting the sexual function could a person enter the highest spiritual state. Asceticism contributed to changes in several early Christian doctrines. My purpose is to explain how these changes took place and to show how they affected the doctrines of the Atonement, marriage, and especially womanhood.

Asceticism seeped into Christianity from at least two (and probably more) sources. First, it came indirectly from a Judaism that had its own ascetic tendencies (e.g., the celibate Essenes at Qumran), and from such Greek-influenced Jews as the philosopher Philo, a contemporary of Jesus and Paul. Second, asceticism entered Christianity directly from early Christian converts who had been trained in Greek philosophy. Many of the earliest Christian converts who wrote to explain and defend Christianity to their Greek-educated friends tried to couch the Christian message in terms their intellectual friends could accept. Because the most profound and basic truths of the gospel (such as a suffering Savior) cannot be reshaped in Greek philosophical terms, these early Christians succeeded not in clarifying and preserving the gospel but in distorting it. It soon became unsophisticated to accept the plain truths surrounding the ministry of the Lord Jesus Christ. President Ezra Taft Benson described the process: "The world shouts louder than the whispering of the Holy Ghost. The reasoning of men overrides the revelations of God, and the proud let go of the iron rod."[2]

In particular, asceticism influenced the thinking of the early Christian authors on the human body, marriage, and procreation. They believed that a better world would arrive if people would quit perpetuating the present fallen order by their acts of reproduction; that is, it would be better to let the material world and fallen man die out. Augustine, for example, declared that there had already been enough souls to fill heaven and that if only all men would follow the Christian example and refrain from the obsolete command to increase and multiply, and rise to the angelic life where there is no more "marrying and giving in marriage," the end of the world would be hastened and the physical cosmos would be transformed into the spiritual New Creation.[3] Thus, the higher spiritual order would come on earth when the old one, the one Eve had precipitated, had ceased.

If man's body was undesirable, then it follows that God could not have a physical body. With the rejection of the material world and man's flesh, these influential ascetics denied God a body. Greek philosophy also insisted that God must be totally unlimited and absolute and that, therefore, he could not be limited to and by a body. In addition, there could be only one such being, because if there were more than one, they could not both be ultimate and absolute.[4] Thus the Christians had a problem when it came to drafting creeds on God's form, because the scriptures plainly taught three divine persons. Therefore, at Nicaea (A.D. 325) the first official council of the Christian Church officially sanctioned the nonmaterial, three-in-one God. A great mystery—a man-made mystery.

An example of a descendant of the Nicene Creed is the Thirty-nine Articles of the Church of England. This creed is typical of orthodox Christian creeds and has a familiar phrase: "There is but one living and true God, everlasting, without body, parts, or passions; ...

and in unity of this Godhead there be three Persons, of one substance, power, and eternity: the Father, the Son, and the Holy Ghost."[5]

The New Testament writers had tried to stanch this flow of ascetical ideas by proscribing ascetic practices while describing a physical and anthropomorphic God. Paul, for example, taught that the fulness of the Godhead dwells in Christ "bodily" (Colossians 2:9) and warned the Colossians to avoid self-abasement and the unsparing abuse of the body (v. 23, according to the Greek translation). Paul taught indeed that man is the *offspring* of God (Acts 17:28). Luke constructed his gospel with specific anti-gnostic elements.[6] The passage in Luke known in textual scholarship as the "bloody sweat" (Luke 22:43–44) provides an example. With the work of various translators, these verses slipped in and out of the manuscripts of Luke because of the physical and passionate Christ that they describe. In another passage Luke emphasized the physical nature of the resurrected Jesus: "Behold my hands and my feet, that it is I myself; handle me, and see; for a spirit hath not flesh and bones, as ye see me have" (Luke 24:39). The apostle John fervently affirmed his personal knowledge of the physical, resurrected Christ (1 John 1:1; see also 1 John 4:3). Nevertheless, the battle for the plain truth was largely lost as the doctrines were changed.

If God was something entirely other than man, that is, if he was a nebulous substance rather than a glorified, resurrected man, then human beings could not become like him: they did not exist on a continuum with God. The diabolical aim was to distance man from God and to scramble men's and women's understandings of their own exalted origin, nature, and destiny.

As alluded to above, the body of the Son of God came under attack as well; one reflection of these Christians failing to accept their own sexuality, and the divine purposes in human sexuality, was that they could not accept God's sexuality either.

Justin Martyr, one of the earliest Christian writers (mid-second century), wrote that Christ had submitted to the conditions of the flesh that included the need for food, drink, and clothing, but one physical function, namely, "discharging the sexual function He did not submit to; for, regarding the desires of the flesh, He accepted some as necessary, while others, which were unnecessary, He did not submit to. For if the flesh were deprived of food, drink, and clothing, it would be destroyed; but being deprived of lawless desire, it suffers no harm. . . . Let not, then, those that are unbelieving marvel, if in the world to come He do away with those acts of our fleshly members which even in this present life are abolished [by some]."[7]

In the third and fourth centuries after Christ, the desert wildernesses of the Middle East filled up with celibate men and women. The practice of asceticism distinguished men and women from the common herd and identified them with the elite. Ascetics found that they had greater prestige as holy persons, and the women, greater autonomy over their lives and their bodies.

Ascetically inclined people reinterpreted and even tampered with various biblical texts to justify themselves and persuade others to their ascetic way of life. Such men as Tatian and Marcion, both writing in the second century after Christ, believed that Christian asceticism was the only way of spiritual deliverance and actually doctored the scriptures to recruit people to a virginal life. Tatian, for example, changed Luke 2:36, which concerns the prophetess Anna who lived with her husband seven years "from her virginity," meaning from the date of her marriage, to say that Anna remained a virgin *with* her husband seven years.

Men like Tatian and Marcion, although undoubtedly neurotic in their hatred of women, nevertheless were amazingly influential in persuading others to the practice of asceticism. Consider this passage from Tertullian, another second-century mainstream Christian writer, on the sin inherent in women: A woman should "go about in humble garb, and rather to affect meanness of appearance, walking about as Eve mourning and repentant, in order that by every garb of penitence she might the more fully expiate that which she derives from Eve,— the ignominy, I mean, of the first sin, and the odium . . . of human perdition. In pains and in anxieties dost thou bear [children], woman . . . ? And do you not know that you are [each] an Eve? The sentence of God on this sex of yours lives in this age: the guilt must of necessity live too. *You* are the devil's gateway: *you* are that unsealer of the [forbidden] tree: *you* are the first deserter of the divine law: *you* are she who persuaded him whom the devil was not valiant enough to attack. *You* destroyed so easily God's image, man. On account of *your* desert—that is, death—even the Son of God had to die."[8] Thus Tertullian lays the blame even for the death of Christ on women. Clearly, misogyny was a factor in the practice of asceticism, based (at least theoretically) on an apostate understanding of the woman's role in the Fall and of the function of the Fall itself.

An apostate group among the earliest Christians called Docetists (from Greek *dokew,* "seems") claimed that Christ only seemed to be physical. This idea invaded mainstream Christianity. Hilary of Poitiers, a fourth-century Christian theologian, wrote of Christ's suffering that "He felt the force of suffering, but without its pain," as if a weapon were to pierce water or fire or air. "The body of Christ by its virtue suffered the violence of punishment, without its consciousness. . . . He had not a nature which could feel pain."[9]

Clearly this immaterial and unfeeling god is a favorite diabolical perversion of truth because of its power to undo the doctrinal fabric. We find it alluded to in the Book of Mormon where Limhi recounts the prophet Abinadi's death as having to do with his teaching that God would come down among the children of men and take upon him flesh and blood and go forth on the earth (Mosiah 7:27–28). And among the apostate Zoramites Alma found the people declaring that God will be a Spirit *forever* (Alma 31:15), denying thereby the eventual incarnation of the Son.

Once God lost his body to this doctrine and could not suffer, the mainspring of the Atonement was effectively removed. A substitute for the Atonement was devised in a practice called penance, apparent in the instructions to women in the passage from Tertullian quoted above. Doing penance meant inflicting self-punishment rather than actually acquiring virtue. Assigned to penitential tasks by a celibate clergy, women and men sat in sackcloth and ashes at the church door, groveled at the feet of the clergy, or abstained from marital relations, sometimes for months. Through these punishments, early

Christians thought they expiated their sins. The practice of penance replaced the true doctrine that Jesus suffered and paid off the law of justice, releasing to women and men a great divine enabling power to pursue the divine nature.

If we had the full text of what constituted the original Bible, we would likely see that changes have been made in at least two main doctrinal areas: first, in the doctrine of the Atonement, especially the accessibility of the grace of Jesus Christ, and, second, in the doctrine of the body of God and the relationship of the Father and the Son; and as corollary, the doctrines of the deification of mankind, eternal marriage, eternal procreation, eternal family, and not least, eternal woman.

These changes in doctrine influenced the whole subsequent development of mainstream Christianity, as we can see today in the enduring practices of penance and celibacy and perhaps in our own possible discomfort in thinking of our Lord in a married relationship; another spin-off is a woman's feeling that she occupies a second-class position in both society and in theology.

Because the doctrine of God's body is so pivotal, it is often the first doctrine that has to be restored before an apostasy can begin to be healed. In the Sacred Grove, Joseph could obviously see that there were two separate, glorified, anthropomorphic beings in the Godhead before either of the Gods had spoken. In a split second Joseph apprehended a truth that had the power to end eighteen hundred years of speculation and pointless philosophizing, centuries of making mysteries out of plainness.

With the restoration of the gospel came a whole complex of related ideas and practices that greatly expanded the knowledge and power of the believers in Jesus Christ. A point I wish to emphasize is that one casualty of an apostasy is diminution of the role of the woman. Not only has the world lost an understanding of her exalted place in time and eternity, but many women themselves have bought the lie. As the true doctrines of marriage and atonement declined through such apostate influences as asceticism, many men and women rejected the traditional roles of womanhood. Women lost their eternal bearings and a knowledge of their eternal status. With the Restoration came a glorious view of marriage and of the exalted position and destiny of women.

President Spencer W. Kimball wrote of woman's equality with man and a divine mission to fulfill, tasks that she accepted in the pre-mortal world: "We had full equality as his spirit children. We have equality as recipients of God's perfected love for each of us. . . . Remember, in the world before we came here, faithful women were given certain assignments while faithful men were foreordained to certain priesthood tasks. While we do not now remember the particulars, this does not alter the glorious reality of what we once agreed to."[10]

When a woman is in full alignment with that which was bestowed on her before this world, her life is filled with power, with clarity, and with a passion to live and fulfill her own unique purposes. This knowledge and power are available to every woman who seeks to know the Lord Jesus Christ and his divine plan for her.

Elder James E. Talmage points ahead to what woman is destined to become, at the same time stirring in us a spiritual knowledge of who the hidden woman is right now:

"Woman occupies a position all her own in the eternal economy of the Creator; and in that position she is as truly superior to man as he to her in his appointed place. Woman shall yet come to her own, exercising her rights and her privileges as a sanctified investiture which none shall dare profane. . . .

"In the restored Church of Jesus Christ, the Holy Priesthood is conferred, as an individual bestowal, upon men only, and this in accordance with Divine requirement. It is not given to woman to exercise the authority of the Priesthood independently; nevertheless, in the sacred endowments . . . , woman shares with man the blessings of the Priesthood. When the frailties and imperfections of mortality are left behind, in the glorified state of the blessed hereafter, husband and wife will administer in their respective stations, seeing and understanding alike, and co-operating to the full in the government of their family kingdom. Then shall woman be recompensed in rich measure for all the injustice that womanhood has endured in mortality. Then shall woman reign by Divine right, a queen in the resplendent realm of her glorified state, even as exalted man shall stand, priest and king unto the Most High God. Mortal eye cannot see nor mind comprehend the beauty, glory, and majesty of a righteous woman made perfect in the celestial kingdom of God."[11]

When a woman accepts in spirit who she was in the premortal world, her beauty of spirit, her valiance, her power, her passion, her great love; and when she accepts who she is destined to become; and when she acknowledges that that gifted, premortal woman is present now in the mortal woman and is seeking expression now in the mortal probation, she will know that she does not have to accept behaviors meant to demean her nor act in self-demeaning ways. Rather she can honor her inner divinity and step forth in her God-given power and birthright to bring to pass much beauty and holiness, and thereby fulfill her own unique earthly mission. Without the restoration of the gospel, you and I would not know nor be empowered to live according to these truths. Thanks to the Lord Jesus Christ, who is restoring who woman really is.

Notes

Revised from "The Restoration of the Doctrines of Marriage and Atonement," in *Women and Christ: Talks Selected from the 1992 Women's Conference Sponsored by Brigham Young University and the Relief Society,* ed. Dawn Hall Anderson, Susette Fletcher Green, and Marie Cornwall (Salt Lake City: Deseret Book Co., 1993), 85–92.

1. For complete documentation of the assertions made in this article and further information on the early Christian apostasy, see the author's work, "The Influence of Asceticism on the Rise of Christian Text, Doctrine, and Practice in the First Two Centuries," dissertation, Brigham Young University, August 1989.
2. Ezra Taft Benson, "Beware of Pride," *Ensign,* May 1989, 5.
3. Augustine, Do Bono Conj. 17–20; De Bono Viduit. 9–11, 23–28. See also Jerome, Eps. 123, 17; Adv Jov. 1, 16, 29; Adv. Helvid. 22.
4. Stephen E. Robinson, "The Great Church Councils," unpublished manuscript in possession of the author.
5. John H. Leith, ed., *Creeds of the Churches,* rev. ed. (Chicago: Aldine Publishing Co., 1973), 266–67.
6. The Christian gnostics were an apostate group that believed that Christ could not have had a physical body.

7. Justin Martyr, *On the Resurrection*, trans. M. Dods, in *The Apostolic Fathers with Justin Martyr and Irenaeus*, reprinted lithograph (Grand Rapids, Mich.: Eerdmans Printing Co., 1985), 1:295.

8. Tertullian, "De Cultu Feminarum [On the Apparel of Women]," trans. S. Thelwall, in *Fathers of the Third Century*, reprint lithograph, ed. Alexander Roberts and James Donaldson (Grand Rapids, Mich.: Eerdmans Printing Co., 1976), 1:1.

9. St. Hilary of Poitiers, *On the Trinity*, 10:23, in *St. Hilary of Poitiers: Select Works*, reprint lithograph, trans. E. W. Watson, L. Pullan, et al., ed. W. Sanday (Grand Rapids, Mich.: Eerdmans Printing Co., 1983), 187–88.

10. Spencer W. Kimball, *The Teachings of Spencer W. Kimball*, ed. Edward L. Kimball (Salt Lake City: Bookcraft, 1982), 315–16.

11. James E. Talmage, "The Eternity of Sex," *Young Women's Journal* 25 (October 1914): 602–3; quoted in *Words of Joseph Smith*, 137, note 4.

.

PRINCIPLES IN PRACTICE

C h a p t e r 1 2

Spiritual Lightening

When we are weighed down by the sorrows of a telestial world, we need some spiritual lightening. Spiritual lightening suggests spiritual principles and powers that can light up our minds and lighten our loads. It may be that it is possible for us to assume too great a burden, to take it all too seriously. This is possible even when the damages in our lives involve sex, drugs, and other heavy sorrows and transgressions—because virtually all damage, innocently incurred or self-inflicted, is ultimately reversible through the Lord Jesus Christ. It is spiritually lightening to realize that in most cases the actual details of the elements of our lives matter less than what we choose to become in the midst of them.

It helps to remember that we came down to a fallen world to experience deliverance from it. The crisis that grips us emotionally may carry a significant message for us. It may be a call from the eternal world to learn something we need to know. Events have a way of conspiring to draw our distracted minds to the voice of the Good Shepherd (Alma 5:37–38), who may whisper through our distress a truth that we have resisted but must now humbly face. "For behold, the Lord hath said: I will not succor my people in the day of their transgression; but I will hedge up their ways that they prosper not; and their doings shall be as a stumbling block before them. . . . But if ye will turn to the Lord with full purpose of heart, and put your trust in him, and serve him with all diligence of mind, if ye do this, he will, according to his own will and pleasure, deliver you out of bondage" (Mosiah 7:29, 33).

This higher perspective teaches that we came down to earth to learn from our own experience the difference between good and evil. To make that judgment we had to come to know—that is, experience—some degree of evil.

The experience with truth that we gain here serves as a small-scale pattern for those very same principles that will be used on a grand scale in eternity—the temporal in the likeness of the spiritual (Moses 6:63; Matthew 25:23). "Whatever principle of intelligence we attain unto in this life, it will rise with us in the resurrection. And if a person gains more knowledge and intelligence in this life through his diligence and obedience than another, he will have so much the advantage in the world to come" (D&C 130:18–19).

The acquiring of spiritual intelligence through experience in a fallen world is the treasure we've come for. Our most valuable gains in intelligence come from embracing the word of God until we can live by his every word (D&C 84:44). Thus the natural gives way to the spiritual, and the intelligence we so obtain will endure forever.

It is also helpful to realize that much of what happens here in the temporal world will pass away into the black hole of eternity and find extinction there: damage we suffered from others will be healed, damage we inflicted on others will be mended, ignorance will give way to the full picture, tears will dry, shattered dreams will find new and eternal expression, lessons we thought we learned too late will find application here and in the world to come. Perhaps this good news is what the Lord is telling an anguishing Joseph Smith when He says: "Know thou, my son, that all these things shall give thee experience, and shall be for thy good. . . . Therefore, hold on thy way, [and] . . . fear not what man can do, for God shall be with you forever and ever" (D&C 122:7, 9).

President Joseph F. Smith taught that in the premortal life we knew much of what lay before us:

"Can we know anything here that we did not know before we

came? . . . I believe that our Savior is the ever-living example to all flesh in all these things. He no doubt possessed a foreknowledge of all the vicissitudes through which he would have to pass in the mortal tabernacle, when the foundations of this earth were laid, 'when the morning stars sang together, and all the sons of God shouted for joy.' . . . And yet, to accomplish the ultimatum of his previous existence, and consummate the grand and glorious object of his being, and the salvation of his infinite brotherhood, he had to come and take upon him flesh. He is our example. . . . If Christ knew beforehand, so did we. But in coming here, we forgot all, that our agency might be free indeed, to choose good or evil."[1]

We came to earth to acquire an essential knowledge that was not fully available to us in the premortal world. We had to come and gain the actual experience of making our way through a plan that was designed to bring across our life paths those experiences we most needed to fulfill the purposes of our mortal probation.

God knew that we would arrive in a fallen world with no memory, no knowledge, and no power to make our way successfully alone. Once in mortality, we would begin to make choices before we had much knowledge, or judgment, or ability to choose right over wrong consistently, and we would inevitably make mistakes. As we grew in a fallen environment we would form wrong opinions, make false assumptions by which we would then govern our lives, and embrace many precepts of men. We would create a veil of unbelief (Ether 4:15) and would, as a result, leave a lot of imperfect products behind us. We would make many choices before we had grasped the significance of even the light that we had. Many would reach an advanced age before really seeing the light. Some would never see it in this life. Provision was made according to the circumstances and true desires of all these people.

So we realize that even though the natural man is wonderfully designed to give way to the spiritual (Mosiah 3:19), he must first, by divine design, experience the errors of the natural mind which would cause him to taste the bitter (2 Nephi 2:15; Moses 6:55). We ourselves may have grown comfortable in the natural mind and may have been slow to give it up. Like Amulek, we might have said: "I did harden my heart, for I was called many times and I would not hear; therefore I knew concerning these things, yet I would not know" (Alma 10:6).

But all the while, life was happening to us; we were making important choices, and these were affecting ourselves and others. Our slowness in changing our minds from natural to spiritual may have caused us to create many sorrowful situations.

We came to earth to acquire an essential knowledge that was not fully available to us in the premortal world. Elder Neal A. Maxwell has written: "Perhaps it helps to emphasize—more than we sometimes do—that our first estate featured learning of a cognitive type, and it was surely a much longer span than that of our second estate, and the tutoring so much better and more direct.

"The second estate, however, is one that emphasizes *experiential learning* through *applying, proving,* and *testing.* We learn cognitively here too, just as a good university examination also teaches even as it tests us. In any event, the books of the first estate are now closed to us, and the present test is, therefore, very real. We have moved, as it were, from first-estate *theory* to second-estate *laboratory.* It is here that our Christlike characteristics are further shaped and our spiritual skills are thus strengthened."[2]

Earth life, then, is something like a laboratory. Our manuals are full of vital instructions, though we may have esteemed them lightly (D&C 84:54–55) and thus find ourselves in a spiritual twilight. Even when we have understood the purposes of life, there is, at times, an undeniable ambiguity in our lives. We do not always understand our lives because "now we see through a glass, darkly" (1 Corinthians 13:12), and "it doth not yet appear what we shall be" (1 John 3:2). We may grope around in this twilight of knowledge, anxious that someone is going to blow up either our experiment or the whole laboratory altogether. It can seem so out of control.

I may see that some of the people around me in the lab are making some dangerous choices. They're not following instructions, they're using the wrong ingredients. I may see that my husband or my teenage son or my married daughter may not be conducting his or her experiments very wisely.

To my dismay, I find that one of my worst fears is that what they are doing makes *me* look bad. (Didn't I teach them how to do their experiments? Won't the world judge me by the way they are doing their lab work? *Aren't* I responsible for how they are doing their work?) Maybe I think that if I can fix them, and they get fixed, I won't have to feel guilty about them anymore. But here I am trying to fix their lives, no matter that I cannot perfectly conduct my own. How can I keep my hands off their experiments? To what degree am I my brother's experiment keeper? As we harbor feelings of confusion, anger, and fear, the spirit of relief eludes us.

We can see on reflection that one of our greatest stressors may be our own pride, our mixing up our own personal value with what another person is doing. Spiritual lightening helps us to straighten that out, answering one of our hardest questions: Where is the line drawn between my responsibility and his or hers? How do I discern between help and interference? How can I have a love pure enough to serve holy purposes rather than my personal agenda? My heart is not pure if I want to control or manipulate for my own ends. With rigorous self-honesty and the fearless shining of spiritual light into the soul, we can cleanse our hearts. We can rise above the twilight (oblivious now to others' opinions), and we can pray: "O Lord, help me to see what appropriate help I can give. Help me to exercise enough faith in thy plan for my loved ones that I can leave them to thee when there is nothing more I can do. Help me to remember that a purpose of their lives is fulfilled in the right to make choices. Help me then to find a serenity that is independent of what another chooses."

One source of relief lies in scriptural examples illustrating this idea of not letting others' choices ruin one's life and health. Here is Alma, weighed down with sorrow, having been cast out of Ammonihah. An angel appears to him and says: "Blessed art thou, Alma; therefore, lift up thy head and rejoice, for thou hast great cause to rejoice; for thou hast been faithful" (Alma 8:15). A similar idea appears in Ether, where Moroni fears that the Gentiles will not have charity for his words. The Lord responds: "If they have not charity it mattereth not unto thee, *thou* hast been faithful" (Ether 12:37).[3] Obviously the extent of our happiness rests primarily on what we do and not on what another chooses to do. We don't need to let another person's choices hold our happiness prisoner. The Spirit can with its rich feelings of inner knowing transcend the otherwise unraveling elements of our lives.

But then a little voice says to our fragile hope: "*But* . . . you *could* have done more; you *shouldn't* have done such and such; you *should* have known; and so forth. You are *not* faithful; *you* caused the problems." That you are even partly to blame for your loved one's trouble may or may not be true. But if I have indeed contributed to a loved one's pain, I can repent. I can make restitution so far as it can be made. After one's heart is truly broken, one might express to another (a mother to a child, or a husband to a wife) some ideas like these: "I am so deeply sorry that I caused you damage. Now, given what I know, I would give anything to undo, to redo, what I did. I wish you could look back on our association together and see me as having done all those things differently. I wish you could recreate the memory and imagine that I acted sensitively, lovingly, patiently, with greater reverence for your agency, your feelings, your needs. I now entrust you to the love of the Lord. I will do all for you that I can, but your troubles and your choices are now between you and your Savior." Our contrition needs to be authentic, deep, and thorough, manifested through our changed behavior and our trustworthiness; but it is not unfeeling to acknowledge that each person's happiness, as well as his salvation, is finally his own choice.

With respect to discerning what to do for others besides changing our own behavior, here also the Lord is our model. We see that he usually avoids doing for us what we can reasonably do for ourselves—what we need to do for ourselves. Much grace is given *after* or *as* we do all that we can do (2 Nephi 25:23). He also often waits for us to ask.

Here, however, are some forms of grace that people greatly need and that can produce unexpected little miracles: empathy, patience, tolerance, forgiveness, listening, kind words, hugs, kisses, smiles, a helping hand, words of encouragement and praise. These are attributes of godliness; they draw the Spirit into our relationships and promote healing.

We remember as well that our purity of heart makes possible that very powerful grace which is received through the Holy Ghost for another in the form of inspired words, spiritual gifts, and so forth. Notice that these have little to do with unsolicited advice giving, or with taking over because we think another can't work out his or her own problems with the Lord. We may indeed be called to intervene in another's life quite directly, but this intervention usually comes most effectively after careful spiritual purification, preparation, and planning.

The scriptures teach that the random nature of the laboratory called mortality is only apparent. The Lord knows everything that is going to happen in the lab before it does. His perfect foreknowledge doesn't mean that he has chosen everything that is going to happen; rather, it means that he has foreseen what we will choose and has put appropriate measures into place so that his great purposes for the individual lives of his children will not be frustrated. Elder Maxwell has written: "By foreseeing, God can plan and His purposes can be fulfilled, but He does this in a way that does not in the least compromise our individual free agency, any more than an able meteorologist causes the weather rather than forecasts it."[4]

The Lord has put into place appropriate compensations, solutions, and healings—specific to damages that will come and specific to the difficult experiences that each of us needs. Many of these solutions and compensations will be realized in this life; others, in the world to come. Of course, agency is always operative, and we will often see people not availing themselves of these divine solutions that are presented again and again through the course of life's experiences. Many will choose not to implement them. But those determined to come to the Lord Jesus Christ will find many, many healings and revelations and opportunities to help them reverse the damages from the mistakes or pain of the past. "The Lord knoweth all things from the beginning; wherefore, he prepareth a way to accomplish all his works among the children of men; for behold, he hath all power unto the fulfilling of all his words" (1 Nephi 9:6).

Elder Boyd K. Packer has said: "Remember that mortal life is a brief moment, for we will live eternally. There will be ample—I almost used the word *time*, but time does not apply here—there will be ample opportunity for all injustices, all inequities to be made right, all loneliness and deprivation compensated, and all worthiness rewarded when we keep the faith. 'If in this life only we have hope in Christ, we are of all men most miserable' (1 Corinthians 15:19). It does not all end with mortal death; it just begins."[5]

One of my major challenges, then, is to accept and work with those things that happen to me in the laboratory that are for my experience and development and blessing—if I can just turn my thoughts and exercise my faith that way (2 Nephi 26:24). The Lord encourages us with these words: "Thou shalt thank the Lord thy God in all things. . . . In nothing doth man offend God, . . . save those who confess not his hand in all things, and obey not his commandments" (D&C 59:7, 21).

My challenge is to see that each tangled relationship and each stressful situation has a personal message for me, to recognize that each is seeking to teach me something I need to know. No matter what specific change it is pointing to, the message always pertains to a better love: "Charity is the pure love of Christ, and it endureth forever; and whoso is found possessed of it at the last day, it shall be well with him" (Moroni 7:47). Then Moroni, quoting his father, Mormon, teaches us how to acquire this quality, indicating that it is a gift that comes under the gentle shaping of the Lord's hand in our lives and action on our hearts: "Wherefore, my beloved brethren, pray unto the Father with all the energy of heart, that ye may be filled with this love, which he hath bestowed upon all who are true followers of his Son, Jesus Christ; that ye may become the sons of God; that when he shall appear we shall be like him, for we shall see him as he is; that we may have this hope; that we may be purified even as he is pure" (Moroni 7:48). Our experiences, then, are designed to draw our attention to what we need to know about the obstacles between ourselves and the pure love of Christ. Thus, whatever it is, the experience comes with a blessing and is one of our best friends.

Since the major purpose of the mortal probation is to learn, not to create unblemished perfection, we can embrace this promise: "And it shall come to pass, that whoso repenteth and is baptized in my name shall be filled; and if he endureth to the end, behold, him will I hold *guiltless* before my Father at that day when I shall stand to judge the world" (3 Nephi 27:16). This scripture suggests that repentance, combined with the power in the Atonement, consumes all the imperfection we've left behind us. The image of the butterfly leaving its lifeless cocoon behind suggests a spiritual metaphor of continuing transformation to greater spiritual beauty. As transformation takes place, the products of the past become biodegradable. I believe the Lord when he says that for the repentant soul, no black list is produced from some divine computer at judgment. He promises: "If the wicked will turn from all his sins that he hath committed, and keep all my statutes, and do that which is lawful and right, he shall surely live, he shall not die. All his transgressions that he hath committed, *they shall not be mentioned unto him*" (Ezekiel 18:21–22).

The details of our lives may from time to time seem to lie in shambles, but all of this is only apparent and it is only temporary. We are seeking the treasure buried in the shambles. As soon as we begin to move toward the light, Amulek promised, *immediately* the great plan of redemption is brought about unto us (Alma 34:31). The Lord told Daniel: "Fear not, Daniel: for from the first day that thou didst set thine heart to understand, and to chasten thyself before thy God, thy words were heard" (Daniel 10:12).

When life's burdens get us down, we can ascend in spirit to a higher reality—a truer one. It is possible during troubled times to feel a sweet spiritual lightening steal into the darkened mind with these impressions: "My [child], peace be unto thy soul" (D&C 121:7); "He that keepeth Israel shall neither slumber nor sleep" (Psalm 121:4). It is possible to feel an irrepressible joy even in the midst of seeming chaos, because, as the Apostle Paul teaches, there is a peace which passes understanding (Philippians 4:7) and a love that passes knowledge (Ephesians 3:19). In the midst of trouble, one really can just smile with recognition and say, like the radio announcer, "This is a test." And we can add, "This is primarily a test; in fact, this is a blessing. The Lord will help me hang on and find the hidden treasure here. I will have achieved fellowship with my Savior, who descended below all things" (D&C 122:8). We remember the Apostle Paul's testimony: "I am persuaded, that neither death, nor life, nor angels, nor principalities, nor powers, nor things present, nor things to come, [n]or height, nor depth, nor any other creature, shall be able to separate us from the love of God, which is in Christ Jesus our Lord" (Romans 8:38–39).

Notes

From *Spiritual Lightening* (Salt Lake City: Bookcraft, 1996), 7–16.

1. *Gospel Doctrine*, 13.
2. Neal A. Maxwell, *All These Things Shall Give Thee Experience* (Salt Lake City: Deseret Book Co., 1979), 19; emphasis in original.
3. The emphasis in this verse and in subsequent scriptural passages in this chapter has been added by the author.
4. Maxwell, *All These Things Shall Give Thee Experience*, 19.
5. Boyd K. Packer, *The Things of the Soul* (Salt Lake City: Bookcraft, 1996), 114.

LOVE AND FEAR

There is no fear in love; but perfect love casteth out fear: because fear hath torment. He that feareth is not made perfect in love" (1 John 4:18).

My remarks here deal rather one-sidedly with the idea that mothers are not entirely responsible for the way their children turn out, for good or for ill. In spite of my weaknesses as a mother, I have wonderful children. Obviously a mother has power to do her children a lot of good. But here I want to discuss mothering that has not been perfect, a kind of mothering we may all have done. How may a woman who fears that she has made serious mistakes with her children, or maybe with anyone she has loved, enter into the rest of the Lord? This chapter concerns parental love and fear.

I come from a family of wonderful people who nevertheless struggled with how to be happy. There were many things we didn't know about living in peace. We mixed our love with fear. My brilliant father suffered from alcoholism, and we all suffered with him. He too had come from a troubled family. He gave up alcohol and made significant strides before he died, but what I experienced in my childhood family seemed to color my life with confusion.

I joined the Church at age nineteen. Though my conversion was real, many of my emotions continued to be out of harmony with gospel teachings, and I didn't know what to do about them. I was not at rest. As a young mother I felt at times that I was only barely keeping my distress from leaking out. But it did leak out. Sometimes I had to struggle to be cheerful at home. I was too often tense with my children, especially as their behavior reflected negatively on me. I was perfectionistic. I was irritable and controlling. But I was also loving, patient, appreciative, happy; I frequently felt the Spirit of the Lord, and I did many parenting things well, but so inconsistently. How could I dissolve this shadow self and become fully resonant with gospel teachings?

Sooner or later the crisis comes for good people who live in ignorance and neglect of spiritual law. The old ways don't work anymore, and it may feel as though the foundations of life are giving way. After some nineteen years of motherhood, my poor overcontrolled children seemed like enemies, I was a workaholic, I was exhausted, and I had no idea where to go for help. I wanted peace.

The day came when I knew I had to have help. During a very tearful, pleading prayer, the Lord spoke to me clearly: "Go home." I made arrangements to fly back to my parents' home, where my father had just finished an intensive rehabilitation program for recovering alcoholics. That drying-out process had gone on before, but this time there was a difference. The program reached out to help all the members of the alcoholic's family, most of whom are troubled and need help until they know who they are, why they feel as they do, and what to do about it.

At home I attended a four-hour orientation program for children and spouses of alcoholics. I read books on being an alcoholic child, I talked to other alcoholics who had become counselors in this rehabilitation program, and I didn't stop praying. I could tell that I was on the verge of a re-creation.

I have found since then that if we don't learn consistent, mature love in our childhood homes, we often struggle to learn it when we become marriage partners and parents. Many people who did not come out of alcoholic families nevertheless suffer from the same kinds of distress as I did. Apparently it doesn't matter what the manifest problem was in the child's family; in a home where a child is emotionally deprived for one reason or another, that child will take some personal emotional confusion into his or her adult life. We may spin our spiritual wheels in trying to make up for childhood's personal losses, looking for compensation in the wrong places and despairing of ever finding it. But the significance of spiritual *rebirth* through Jesus Christ is that we can mature spiritually under his parenting and receive healing compensation for these childhood deprivations.

Three emotions that often grow all out of proportion in the emotionally deprived child are fear, guilt, and anger. The fear grows out of the child's awareness of the uncontrollable nature of her fearful environment, of overwhelming negative forces around her. Her guilt, her profound feelings of inadequacy, intensify when she is not able to make the family situation better, when she is unable to put right what is wrong, either in the environment or in another person, no matter how hard she tries to be good. If only she could try harder or be better, she could correct what is wrong, she thinks. She may carry this guilt all her life, not knowing where it comes from but just always feeling guilty. She often feels too sorry for something she has done that was really not all that serious. Her anger comes from her frustration, perceived deprivation, and the resultant self-pity. She has picked up an anger habit and doesn't know how much trouble it is causing her.

A fourth problem often follows in the wake of the big three: the need to control others and manipulate events in order to feel secure in her own world, to hold her world together—to *make* happen what she wants to happen. Here is ignorance of the reality of God's power to hold the universe together. She thinks she has to run everything. She may enter adulthood with an illusion of power and a sense of authority to put other people right, though she has had little success at it. She thinks that all she has to do is try harder, be worthier, and then she can change, perfect, and save other people. But she is in the dark about what really needs changing.

As my own part in my distress began to dawn on me, I thought I would drown in guilt. I wanted to fix all the people that I had affected so negatively. But I learned that I had to focus on getting well and leave off trying to cure anyone around me. We seldom see how much we are a key part of a negative relationship pattern; thus, I recognized that as I healed, many of those around me might indeed get better too. I have learned that it is a true principle that I needed to fix myself before I could begin to be truly helpful to anyone else; I began to understand motes and beams (Matthew 7:5).

I learned the serenity prayer: "God grant me the serenity to accept the things I cannot change, courage to change the things I can, and wisdom to know the difference."[1] I used to think that if I were worthy enough and worked hard enough, and exercised enough anxiety (which is not the same thing as faith), I could change anything. But I learned that my power and my control were illusions. I learned that to survive emotionally I had to turn my life over to the care of that tender Heavenly Father who was really in charge. I learned that it was my own spiritual superficiality that was making me sick, and that only profound repentance, that real change of heart, would ultimately heal me. My Savior was much closer than I had dreamed and was willing to take over the direction of my life: "I am the vine, ye are the branches: He that abideth in me, and I in him, the same bringeth forth much fruit: for without me ye can do nothing" (John 15:5). My brain knew these truths, but to my heart they were new insights.

The crisis I'd precipitated with my violations of eternal law also provided moments of blazing insight. As the old foundations crumbled, I felt terribly vulnerable. I learned that humility and prayer and flexibility are keys to passing through this corridor of healthy change while we experiment with truer ways of dealing with life. I learned things, as Moses did, that I never before had supposed (Moses 1:10). Godly knowledge, lovingly imparted, began deep healing for me, gave me tools to live by and new ways to understand the gospel. "If thou shalt ask, thou shalt receive revelation upon revelation, knowledge upon knowledge, that thou mayest know the mysteries and peaceable things—that which bringeth joy, that which bringeth life eternal" (D&C 42:61). Here, for example, are some peaceable things I learned:

First, each parent brings to her parenting personal weaknesses which will provide opposition for her children. Although we do not want to be the source of even a part of our children's opposition, most of us are, in one way or another. That is a sobering observation, though not surprising. But we remember that God knew ahead of time the failings, the confusions, misunderstandings, weaknesses, and spiritual infirmities of each of his children. With this knowledge he prepared our personal gospel plan. He allowed us the experiences of mortality, but at the same time provided certain compensations and blessings and talents which would present themselves as we struggled with the opposition along life's path.

Most of us parents bring out of our childhood some consequences of our own parents' spiritual infirmities. Of course, we are only describing the conditions of a fallen world: it is a world of weakness and infirmity, but it is also a divinely sanctioned learning environment for this stage in our progress. Perhaps one of the most important views of life to embrace is that this life is a series of tutorials designed to give us experience, to develop the divine nature, and to send us to the Lord Jesus Christ, the Master Teacher and Keeper of Grace. So it has seemed to me that parenthood may be designed at least as much for parents as for children.

Again, I do not want to diminish the fact that a mother can do much necessary good for her child, but why do some of us learn valuable parenting lessons after it seems mostly too late to incorporate them? Perhaps because it is never too late, really, in the eternal scheme of things. It seems that when Mother gets better, everyone in the family gets better—no matter how old the children are.

Second, children's personalities are not initially created by their parents; they have their own unique personality chemistry, which they bring from the premortal life. Much of this personality formation exists beyond the parents' control. What the child chooses to do is a unique reaction resulting from his own personality, his agency, and his environment. Thus some children out of the seemingly best environments have serious problems, while others out of very deprived environments may show amazing maturity and resilience.

As children grow, they gradually become responsible for what they do with the opposition in their lives. I mentioned earlier that although children are born to imperfect or even emotionally sick parents, they have access to divine resources to help them overcome and benefit from their imperfect environment. For example, they may have counterbalancing spiritual gifts, character traits, special blessings, or solutions that will present themselves on life's path again and again as they grow. The children then gradually become responsible to partake of those solutions that the Savior offers them individually.

As solutions become apparent, we as parents may find ourselves in a difficult situation: We think we have produced trouble in a child's life. We now think we know what the solution is. We want to undo the trouble. We offer our best solution. We may be right or we may be wrong, but we want that child to apply the solutions we think he needs—right now. It is very frustrating when a child will not. We find we cannot undo the trouble as quickly as we want it undone. We learn that no child of any age can be forced to accept solutions. The child may even prefer the trouble to the solution. Parents can pray for, but must also wait for, readiness in a child. We learn divine patience, coming to understand our eternally patient Savior more and more.

In the meantime it is helpful to remember that as a child makes mistakes and lives with the consequences, he is gaining experience. Even if he seems to be living in terrible danger of one kind or another, he exists in a universe overseen by an omnipotent and loving God on whom nothing is lost.

One of our problems is that we feel so responsible. Latter-day Saints are very responsible people. Perhaps some of us assume more responsibility than is ours; we may even try to assume God's responsibility. We may think we bear the whole load of another's salvation and forget that every person is a child of God and has his own personal Savior. I learned that I am not my loved ones' Savior. Love easily turns into unrighteous dominion, into interference with a plan that must unfold between a person and his Savior. When love tries to control, it is not love, but fear. I am trying to resist the temptation to control the choices my loved ones make.

Third, no mother, not even a sick mother, can deprive her child of the celestial kingdom. A person's exaltation is ultimately his own responsibility, not his mother's. Women may give lip service to the principle that we are finally responsible only for our own salvation, but in the same moment we may be filled with the terrible fear that we have failed our children irreparably. There seems no restitution we can make. Fortunately, we are wrong. Restitution comes in the mother's healing process. We must avoid thinking that our children's salvation rests solely on us. We are but one component in a multi-faceted plan.

Fourth, with all our hearts we do not want to be the cause of suffering for our child; but where suffering is inevitable, we find solace in the knowledge that all people must suffer in order to learn life's greatest teachings. People have a right, even a need, to suffer. Suffering will often propel us to God, where we learn life's most precious mysteries. We may forever try to get between our children and their painful experiences, whether we created the pain or not. But there are some lessons that only pain can teach, and for us to try to remove all pain—especially the pain of consequences for sin—is beyond our stewardship and control. We may thwart the plan of God by trying to remove consequences.

So we learn to be wiser than we have been and to go in a direction that we had never before thought to go: we let go and let God. We learn patience and faith in the unfolding of an individual plan for each of our children.

Fifth, many mothers have had children who have experimented with a few things. These mothers have agonized. They have stormed the gates of heaven. They have cajoled. They have taught until the children were sick of Mother and the gospel. They have reproached, they have pled, they have threatened. These mothers can learn: love must be governed by intelligence, patience, and faith, not anxious emotion. Jacob says, "I will unfold this mystery unto you; if I do not, by any means, get shaken from my firmness in the Spirit, and stumble because of my over anxiety for you" (Jacob 4:18). Overanxiety can actually block the Spirit. Maybe that's why the Savior of the world reminds us, "Be still, and know that I am God" (Psalm 46:10).

It is very hard to let go of that agonizing knot in the intestines that urges us to take all the blame and to beat ourselves mercilessly. But if we are going to learn godliness we must redirect the energy that we're giving to sick anxiety over another's choices. We do not perfect others with our fear, or even with our love; rather, we perfect the principle of love in ourselves.

To our children there may be something unwholesome about our love if they feel too much anxiety in it. They may feel they are being manipulated by our so-called love. They don't receive it as love, and perhaps they are right. Rather than love, it may be more like fear, or hurt pride, or anger over failure to control. They must see that our happiness is not dependent on their conversion. We must take care not to try to manipulate them with our unhappiness.

Rather, as God is at peace in spite of our choices, so may we be at peace in spite of anyone else's choices. Alma tells his wayward son, Corianton, that the nature of God is a state of happiness (Alma 41:11). One's nature can be independent of what others do. "All truth is independent in that sphere in which God has placed it, to act for itself, as all intelligence also" (D&C 93:30).

Shall we destroy ourselves with grief and self-reproach? Does God destroy himself over the choices of his children? Surely not; otherwise the celestial kingdom would have no peace for him or us. We do not struggle to gain our exaltation just so we can suffer on a grander scale.

Do we love our children more if we allow our grief over them to destroy us? Do we love them less if we allow them their agency and establish serenity in our own bosoms? Can we actually detach ourselves from the tyranny of our grief? If we are going to be gods, we must realize that our loved ones are independent agents; we must get used to people rejecting our best gifts.

Thinking about serenity, or the Lord's rest, consider these scriptures:

Alma 33:23: "And then may God grant unto you that your burdens may be light, through the joy of his Son. And even all this can ye do if ye will."

If we will cast our burdens on the Lord, as he invites us to do, and not cling like martyrs to them, we will find relief. I ask myself, "Why am I hurting myself?" God doesn't ask me to. Just the opposite. Perhaps I have liked thinking that my anxiety was a measure of my spiritual sensitivity or my love, but I have had to admit that really it was either neurosis or lack of faith, or both.

Moroni 7:3: "Wherefore, I would speak unto you that are of the church, that are the peaceable followers of Christ, and that have obtained a sufficient hope by which ye can enter into the rest of the Lord, from this time henceforth until ye shall rest with him in heaven."

Mormon observes that the Church members to whom he is speaking have already entered into the Lord's rest. Did their right to God's rest mean that they didn't have difficult children or errant loved ones, or that they had never made mistakes? I am learning that my rest in God is not dependent on others' choices, not even my children's. It is dependent on my own repentance and my implementation of saner principles.

D&C 6:36: "Look unto me in every thought; doubt not, fear not." "Pray always, and I will pour out my Spirit upon you, and great shall be your blessing. . . . Yea, come unto me thy Savior" (D&C 19:38, 41).

Prayer has brought many a wiser parent, and then even a wayward child, home again (Mosiah 27:22).

God lives, and our attentive Savior lives, who is willing to take loving charge in the life of each of us. There are divine solutions to each of our most distressing dilemmas. There are principles to learn and to work by. There is restitution. We may indeed cast out fear, and love with serenity.

Notes

From *Spiritual Lightening* (Salt Lake City: Bookcraft, 1996), 65–74.

1. "The Serenity Prayer," in *One Day at a Time in Al-Anon* (New York: Al-Anon Family Group Headquarters, Inc., 1983), last page (unnumbered).

LIVING THE SPIRIT OF
AT-ONE-MENT

The Lord Jesus Christ's atonement for you and me made possible an at-one-ment society. The spirit of the Lord's at-one-ment is always seeking access to our relationships, but this spirit can seem very elusive in our personal and workaday worlds. The world we live in seems to have little relationship to a Zion society. Nevertheless, the powers in the atonement apply to our lives and are accessible to you and me right now. These powers have implications for every relationship we have and in every combination of people in which we find ourselves. They have implications for what we think and say, what we do, and how we feel. These powers may hold the secret to making right relationships endure and may help us to know what is wrong with potentially good relationships that are going wrong.

What are the principles on which a Zion society or a community of at-one-ment is established? First, we need a little background on the word *atonement*. *Atonement,* literally *at-one-ment,* is a word introduced into English in 1526 by William Tyndale as he translated the Greek New Testament into English; specifically, he created the word *at-one-ment* to translate the Greek word (*katalge*), which means *reconciliation* or *to come back into a relationship after a period of estrangement.* This word points to what has happened to man—he has fallen from a relationship, even many relationships, and from a knowledge of the oneness of the premortal children. The scriptures tell us that man came from a heavenly society and fell, by his birth, into a state of spiritual death (Helaman 14:16), alienated from his Heavenly Father by the nature of the Fall. Christ wrought the atonement to restore us to the heavenly society. So we might say that the word rendered *atonement* by the early biblical translators could have been more accurately rendered *re-at-one-ment* or *reunion*. Christ wrought the great Reunion.

Brigham Young, referring to the need for the reintegration of mankind, says that the Prophet Joseph showed him in a vision the premortal organization of the human family, then said: "'Be sure to tell the people to keep the spirit of the Lord; and if they will, they will find themselves just as they were organized by our Father in Heaven before they came into the world. Our Father in Heaven organized the human family, but they are all disorganized and in great confusion.'

"Joseph then showed me the pattern, how they were in the beginning. This I cannot describe, but I saw it, and saw where the Priesthood had been taken from the earth and how it must be joined together, so that there would be a perfect chain from Father Adam to his latest posterity."[1]

The work of Christ is the work of reintegration. Scriptural uses of *atonement* or at-one-ment suggest that Christ intends to bring us to oneness in heaven and to that social harmony that we experienced before the world was; a harmony, in fact, that still continues in heaven, and into which you and I seek to be readmitted; a society that the Prophet Joseph described as "sociality . . . coupled with eternal glory" (D&C 130:2). Preparation for that society is our goal.

We have imprinted on our spirits the at-one-ment experiences from the premortal heavenly community, even though we do not remember the details. Parley P. Pratt wrote that after God's spirit children were born, they were "matured in the heavenly mansions, trained in the school of love in the family circle, and amid the most tender embraces of parental and fraternal affection."[2] Perhaps in this life we unconsciously measure every relationship against what our spirits remember. Perhaps we suffer at some level when we do not experience what we once knew.

Our present human sociality is only a shadow of that perfection that existed in the premortal world, where the heavenly society functioned by eternal principles of love. Through the gospel, those principles are available to us here. It is one of the purposes of the gospel to help us implement those principles here in order to reclaim our place in heaven. Therefore, we might observe: things must be done on earth as they are done in heaven so that that which is earthly may be made heavenly. That which does not try to be heavenly must remain telestial and cannot be made heavenly or celestial.

At-one-ment is the condition in which heavenly beings live. If we want to live there with them, we can practice here and now the manner of emotional and spiritual life that they live. *This* life is the time for men to prepare to meet God (Alma 34:32). We want to live the spirit of at-one-ment in whatever ways we can on the earth. We may feel the Holy Spirit already working on us to live in at-one-ment, but we don't always know how to respond to the promptings.

Oneness is a comprehensive principle and of primary importance with the Lord. Speaking of those with whom he will one day "drink of the fruit of the vine . . . on the earth," he referred to those through whom he would "*gather together in one* all things, both which are in heaven, and which are on earth; and also . . . all those whom my Father hath given me out of the world" (D&C 27:5, 13–14).[3]

Scripture abounds with references to being *one,* to crying with one voice to the Lord, to being of one heart and one mind. These references to oneness in the experiences of the people reveal that the spirit of at-one-ment is working upon the people to draw them to a higher spiritual plane, where the powers of heaven can be manifested to a greater degree. The scriptures use many at-one-ment words and phrases: oneness, in one, unity, united order, having things in common, gathering (versus scattering), equal, cleave, seal, welding link, embrace, consecration, marriage, restoration, resurrection. We do the work of at-one-ment in the temple when we seal our ancestors and posterity to us in great family chains. *At-one-ment* is another word for sealing.

Effective prayer in the temple requires feelings of love. The temple endowment itself is a progressive sealing of ourselves to the Lord until we are clasped in the arms of Jesus (Mormon 5:11). Considering how comprehensive the Lord's at-one-ment work is, we can see that it is physical, emotional, and spiritual.

But any work that involves gathering Saints together is at-one-ment work. The gathering of Israel by the missionaries is one example, as is organizing converts into the wards and stakes of Zion. Home teaching and visiting teaching all partake of the purposes of at-one-ment in Zion. Family home evening is at-one-ment work.

We want the spirit of the at-one-ment in our personal relationships, since the ordinances we receive in the temple are inextricably linked with the principles of love. We learn there that spirituality and spiritual gifts and sealings cannot be separated from loving behavior and feelings.

Consider the Zion society that resulted from the Lord's visit to the Nephites: "And they had *all things common* among them; therefore there were not rich and poor, bond and free, but they were all made free, and partakers of the heavenly gift. . . . And it came to pass that there was *no contention* in the land, because of *the love of God* which did dwell in the hearts of the people. And there were no envyings, nor strifes, nor tumults, nor whoredoms, nor lyings, nor murders, nor any manner of lasciviousness; and surely there could not be a happier people among all the people who had been created by the hand of God.

" . . . They were *in one,* the children of Christ and heirs to the kingdom of God. And how *blessed* were they!" (4 Nephi 1:3, 15–18).

This description of the at-one-ment among the children of Christ illuminates compelling possibilities. How do we get to the heavenly, Zion condition described in 4 Nephi? How do we learn how to feel the spirit of the at-one-ment? How do we bridge the gap between where we may now perceive ourselves to be spiritually and where we want to be? Will the Lord do something magical to us to make us ready for his coming, to make us ready to build Zion, to enter at last into the kingdom of God? How do we get there from here?

Many have gotten there and will get there from here (see, for example, Alma 13:12). A group of Latter-day Saints will yet prepare to build Zion and they will live in it. And one day the Savior will appear, bringing with him that heavenly society that has been engaged in the principles of at-one-ment for eons in the celestial world. When they come to the earth, they will find a society that has also practiced these same principles of union, a people that will have perfected themselves by living these principles. We will join societies and be one with them—and without doubt, they will teach us yet more about union and love.

So what can we do to ensure that we are prepared to participate in such a society? We realize that one of our purposes in life is to learn the principles that govern Zion societies and to take them with us into the kingdom. We knew these principles in the premortal world, but here we may have lost our sense of connection.

The opposition to at-one-ment can seem formidable when we realize how easily in times past we may have traded the spirit of at-one-ment for disturbance. We see what a challenge it might be for us to live in a Zion or heavenly condition where everyone will have learned, by desire and practice, to prefer the spirit of at-one-ment with each other to that of conflict or disturbance.

What is the nature of the negative energy that leads to conflict around us? It has something to do with what we think we need from other people in order to be happy. So we manifest the opposite of at-one-ment (disintegration, disconnection), by our un-peace, restlessness, imposing our own will; by criticism, anger, irritability, selfishness, failure to forgive, failure to revere another's agency, retaliation, moodiness, fear, worry—simply forgetting to have faith in the Lord Jesus Christ. All of these we have probably all experimented with to learn bitter and sweet. These are ways most of us act until we learn that there is a better way.

We all feel negative emotions, and sometimes they need to be expressed—carefully. But even when these negative-energy emotions are fully justified, they can constitute a spiritual burden. Our bad temper and bad moods become a form of abuse for us and those around us.

It may take more than a little humility to accept this truth. Perhaps we have not fully processed the idea that the Spirit flourishes in the peaceful atmosphere of at-one-ment. We may not have realized the spiritual value of inner peace (Mosiah 4:13). Mormon speaks to the followers of Christ, whom he recognizes because of their peaceable walk with men, people who have entered into the rest of the Lord (Moroni 7:3–4).

As I have watched myself and others, it is sobering to realize how readily we trade inner peace for something less, for some sort of upset; how readily we take offense and then escalate the disturbance around us—in home or office or even church. How easily we have unsatisfied expectations of how others should treat us or what they should be doing for us—and we grow cold or irritable to retaliate for this real or imagined slight. How eagerly we may insist on being right at the expense of precious relationships. Thus keeping the water rippling around us with negative energy, we are often not still and at rest in the principles of tolerance and love, of overlooking, of letting go, of forgiving.

I find that when I am not at peace inside, I make trouble around me. I even look for trouble, picking at this, complaining at that, practicing abuse. I may yield to self-pity that causes me to withdraw, licking my wounds, waiting for someone to put right what is really my responsibility to correct inside myself. It may be that self-pity is a sin because it violates the spirit of at-one-ment and faith. It sees life through the eyes of a victim rather than from the perspective of a disciple of Christ. I have asked myself, How long could I last in Zion? How long would it be before I single-handedly dismantled Zion?

Maybe I have thought that at the last judgment someone would wave a magic priesthood wand over me and I would suddenly acquire a heavenly personality. But it's clear now that the Lord expects me to practice at-one-ment here and to involve him in helping me in these kinds of personal challenges until the heavenly personality becomes mine.

A Zion society is the product of the personal choice of every person in it; it is also a function of the enabling power of the Lord Jesus Christ that shapes hearts to be like his. But first it begins with an individual choice, independent of others' choices for something less. I have come to know that in any moment what I send out is my choice, and I can't blame it on a situation or on another person. "And now remember, remember, my brethren, that whosoever perisheth, perisheth unto himself; and whosoever doeth iniquity, doeth it unto himself; for behold, ye are free; ye are permitted to act for yourselves; for behold, God hath given unto you a knowledge and he hath made you free. He hath given unto you that ye might know good from evil, and he hath given unto you that ye might choose life or death; and ye can do good and be restored unto that which is good, or have that which is good restored unto you; or ye can do evil, and have that which is evil restored unto you" (Helaman 14:30–31).

The power of evil opposes at-one-ment. We mortals are not alone on this planet. The Apostle Paul wrote: "Put on the whole armour of God, that ye may be able to stand against the wiles of the devil. For we wrestle not against flesh and blood, but against principalities, against powers, against the rulers of the darkness of this world" (Ephesians 6:11–12). Brigham Young said in connection with the subtle works of Satan: "There are thousands of plans which the enemy of all righteousness employs to decoy the hearts of the people away from righteousness."[4] The Prophet Joseph focused the idea: "The policy of the wicked spirit is to *separate* what God has joined together, and unite what He [God] has separated, which the devil has succeeded in doing to admiration in the present state of society."[5]

Satan seeks to rend the Saints' relationships—their marriages, their family feeling, their ward associations, their business connections—so that the powers of Zion cannot be established. We forget in the very moment that Satan and his followers promote contention (3 Nephi 11:29) by stirring around in the pride in the Saints' hearts. Consider these eye-opening scriptures:

—"But, O my people, beware lest there shall arise *contentions* among you, and ye list to obey the evil spirit" (Mosiah 2:32).

—"And [Alma] commanded them that there should be no *contention* one with another, but that they should look forward . . . having their hearts knit together in unity and in love" (Mosiah 18:21).

—"But behold, this was a critical time for such *contention*s to be among the people of Nephi. . . . It was [Moroni's] first care to put an end to such *contentions* and *dissension*s among the people; for behold, this had been hitherto a cause of all their destruction" (Alma 51:9, 16).

—"And many more things did the people imagine up in their hearts, which were foolish and vain; and they were much disturbed, for Satan did stir them up to do iniquity continually; yea, he did go about spreading rumors and *contentions* . . . that he might *harden the hearts* of the people against that which was good" (Helaman 16:22).

—"For verily, verily I say unto you, he that hath the spirit of *contention* is not of me, but is of the devil, who is the *father of contention,* and he stirreth up the hearts of men to contend with anger, one with another" (3 Nephi 11:29).

—"I . . . establish my gospel, that there may not be so much *contention;* yea, Satan doth stir up the hearts of the people to contention" (D&C 10:63).

—"There were jarrings, and *contentions,* and envyings, and strifes, and lustful and covetous desires among them [the Saints]; therefore by these things *they polluted their inheritances.* They were slow to hearken unto the voice of the Lord their God; therefore, the Lord their God is slow to hearken unto their prayers, to answer them in the day of their trouble" (D&C 101:6–7).

—"Cease to *contend* one with another; cease to speak evil one of another . . . and let your words tend to edifying one another" (D&C 136:23–24).

—"And now my beloved brethren, I would exhort you to have patience, and that ye bear with all manner of afflictions; that ye *do not revile* against those who do cast you out . . . , lest ye become sinners like unto them; but that ye have patience, and bear with those afflictions, with a firm hope that ye shall one day rest from all your afflictions" (Alma 34:40–41).

"We must attend to the ordinance of washing of feet. . . . It is calculated to unite our hearts, that we may be one in feeling and sentiment, and that our faith may be strong, so that *Satan cannot overthrow* us, nor have any *power* over us here. . . . Do not watch for iniquity in each other, if you do you will not get an endowment, for God will not bestow it on such. But if we are faithful, and live by every word that proceeds forth from the mouth of God, I will venture to prophesy that we shall get a blessing that will be worth remembering, if we should live as long as John the Revelator."[6]

Having discerned Satan, then, we can thwart the evil spirit in many ways. You and I have been endowed with the divine power to generate positive energy—spiritually, mentally, physically—by carefully choosing attitudes, actions, and words to promote a spiritually nourishing environment. Each of us has that power. We can choose to generate positive, spiritual energy, with which the Spirit connects, and which he magnifies, creating daily miracles in relationships. Thus we learn to work as the Savior works and to become as he is, even as we walk this life.

But we may have many misconceptions about how to be happy and how to establish relationships of at-one-ment with others. We may think these relationships have to be ideal; we may think that the people around us have to be ideal, that they have to feel and think the way we do in order to be happy, or that we have to think as they do in order to have the spirit of at-one-ment between us. We may feel that many of the people around us do not value what we do, do not meet our hopes and dreams, and we may despair that we will ever experience at-one-ment with some of the people God has put into our lives.

Here indeed is the reality of telestial living—nearly every day someone will do to us one or more of the following: belittle, be insensitive to needs, show indifference, make us feel insecure, humiliate, frighten, abuse, inconvenience, demand, criticize, disappoint, lie, hurt, betray, try to seduce, misunderstand, resent, threaten, attack (verbally or physically).

So what shall we do about all that? What we do is important. We can't, before God, blame our response on what others do to us. In fact, our earth life was designed to make those abrasive experiences possible, as a sort of laboratory in which we could work out our salvation. It is in these daily abrasions that we find the imperfections in our own souls.

Maybe one of the purposes for such experiences can be answered with this question: How shall we ever learn Christlike love unless we have a chance to practice it in the face of the opposites? Every disrupted relationship, whether in our own home or within a particular group or community, is a chance to forge the divine nature in ourselves and prepare for that endless state of happiness.

It would appear that all the people in our lives are there for important reasons. We stand in a sacred relationship to them because we and they cannot be made perfect without each other (D&C 128:18—the Prophet speaks of a *welding link* that must be established). Nevertheless, we remember that seldom are they given to us to satisfy us. Rather, they are given to us to make possible a much greater love than we would have been capable of in a situation where everybody agreed with us, everybody loved us, everybody saw everything the way we do. These abrasive people in our lives are friends in disguise. They are there to teach us to perfect love in ourselves, not to perfect them. We don't need ideal relationships in order to be happy; we can live happily with less than the ideal because each precious relationship can be made more tender and sweet and can be enriched with that spirit of at-one-ment that changes everything around us.

All of us have experienced or are now experiencing troubled relationships. I know from my own experience that miracles often happen in troubled relationships. I grew up in a troubled family; each individual in that family was and is a good person. They were good people with very little understanding in those early years of how to be happy. But each of us has come a long way since then.

In a troubled family one may learn a number of counterproductive behaviors: to try to control others; to be critical in order to feel more secure in one's own self-righteousness; to require satisfaction from others' behavior; to use anger as manipulation; to be very self-assertive; to try to prove oneself right in every situation; to make trouble by letting people know the various ways in which they are not meeting one's expectations; to get even by using irritability, cold silences, or not-so subtle barbed words; to nag people and try to talk them into things. The people we treat in these ways come to feel they are our enemies. Often we create these enemies within our own family circle. The results of these behaviors are that we experience a lot of unfocused fear and anger, as well as tendencies to depression, guilt, and feelings that life is meaningless.

I didn't know there was really anything wrong with me as I practiced some of these ways of treating people, but I did know I didn't feel good. I did not see a relationship between the way I treated other people and the way I felt inside: I thought that what they were doing made me unhappy; but actually it was how I was treating them that made me unhappy. Is it possible that much of the emotional pain we have comes not from the love we weren't given in the past, but from the love we ourselves aren't giving in the present?

We can get mixed up. We know that we are required to do all that we appropriately can to promote the well-being of those the Lord has entrusted to our care. But when older children and adults behave in ways that are very distressing to us, it is easy to become involved in ways of acting that do not help them or ourselves. We may be so emotionally entangled that we think obsessively about what the other person is doing, and this involvement only keeps us in turmoil. Sometimes our over-involvement is a blend of resentment, self-pity, and guilt. But we who wish to enter into at-one-ment must first learn a special detachment from others whose behavior we can't control. Detaching ourselves emotionally, ceasing to manipulate the other person's life, letting that person take responsibility for his or her own behavior—this release frees us from soul-sickening stress.

This detachment does not imply that we withdraw our love and compassion or any appropriate help. It means that we can turn our attention to the things we have neglected, the things that truly are our concern. Our detachment produces inner serenity as we take full responsibility for what we do, repenting and correcting ourselves as necessary, and giving others responsibility for what they do. This kind of detachment is essential to any healthy relationship.

It is not by accident that the Book of Mormon describes some dysfunctional families and troubled relationships. Nephi, for example, lived with abusive older brothers, receiving verbal and physical abuse from those who should have been his protectors and nurturers. Although on at least one occasion he is able to frankly forgive them (1 Nephi 7:21), he later reacts with anger, only he has turned his anger inward—a common source of depression. He sees that, although his anger is 100 percent justified by telestial standards, for his own spiritual well-being he must let it go and turn to the Lord for release. "Why am I angry because of mine enemy? Awake my soul! No longer droop in sin. Rejoice, O my heart, and give place no more for the enemy of my soul. Do not anger again. . . . Do not slacken my strength. . . . Rejoice, O my heart, and cry unto the Lord. . . . Wilt thou make me that I may shake at the appearance of sin?" (2 Nephi 4:27–31). Nephi teaches this powerful principle: Our happiness depends on what we do now, not on what was done to us.

Certainly we are not talking here about submitting to serious abuse. Forgiving people, acting kindly toward them, doesn't necessarily mean letting them abuse us. Sometimes relationships have to be severed to keep one of the parties from being destroyed. In Nephi's case, the Lord finally took him out of Laman and Lemuel's presence (2 Nephi 5). But Nephi waited on the Lord, teaching us that revelation is indispensable to relationship work.

Each relationship has a partly hidden history. What we do not know or remember is what we covenanted to do in the premortal world with respect to a particular relationship here. In some cases the Lord will take us out of a relationship, or counsel us to take ourselves out, but very often he will set about to work a small series of miracles in the relationship so that the spirit of at-one-ment can flourish. He is preparing us to live in a celestial society; therefore, it seems that usually he wants us to mend rather than sever relationships.

With respect to mending relationship trouble there is a stunning insight to process: if there is trouble, *each* person in the relationship has contributed to some part of the problem; therefore, *each* one has something to repent of as well as something to learn from the experience—and also some new choices to make with respect to the relationship. We can't begin to heal a relationship until we've acknowledged our own part in its dysfunction.

Truly we receive what we send out: "For that which ye do send out shall return unto you again, and be restored" (Alma 41:15). If we don't like what we're getting in a particular relationship, we may need to check out what we're sending into that relationship.

Each relationship has a "measure of creation." If we have neglected to fill that measure with our own part, our spirit will smite us, usually at some lower level of consciousness. When our spirit is smiting us, we often start smiting others and projecting our own guilt onto them. We find all kinds of things wrong with them. Our own unconscious guilt causes us to be unduly affected by others. It is when we stop sinning against others with our neglect and unrighteous expectations that we come to view our relationships to others differently. We are less concerned with what others are doing; in fact, a lot of the old ego issues and offenses just come to seem irrelevant.

A quiet conscience seems to free up a lot of energy, as well as a desire to bless others with no other motive than to love. A person with a quiet conscience, one who is increasingly conscious of a sweet, divine presence, does not need others to satisfy him because he doesn't need much from others to be happy. He knows that his security and wholeness do not depend on what others think, say, or do; rather, these things depend on what he himself chooses to do. His choices either align him with the eternal laws of harmony or alienate him from himself and others.

We always have a choice. For example, when someone trespasses against me, I may feel a negative ripple through my being and I face a moment of decision: shall I return the assault, making a poor situation worse? Or shall I neutralize this assault by returning kindness, good for evil? Paul says, "Be not overcome of evil, but overcome evil with good" (Romans 12:21). "Love your enemies, bless them that curse you, do good to them that hate you, and pray for them which despitefully use you, and persecute you; that ye may be the children of your Father which is in heaven" (Matthew 5:44–45; for additional insight see D&C 64:8–10; JST, Luke 6:29–30; Matthew 5:39–40).

The Prophet Joseph's life reflects the relationship between the principles of forgiveness and the gifts of the Spirit. David Whitmer said: "He [Joseph Smith] was a religious and straightforward man. . . . He had to trust in God. He could not translate unless he was humble and possessed the right feelings towards everyone. To illustrate so you can see: One morning when he was getting ready to continue the translation, something went wrong about the house and he was put out about it. Something that Emma, his wife, had done. Oliver and I went upstairs and Joseph came up soon after to continue the translation but he could not do anything. He could not translate a single syllable. He went downstairs, out into the orchard, and made supplication to the Lord; was gone about an hour—came back to the house, and asked Emma's forgiveness and then came upstairs where we were and then the translation went on all right. He could do nothing save he was humble and faithful."[7]

These are the spiritual sensitivities by which Zion is established.

Brigham Young taught about being worthy to associate with angels—for that is indeed what we are seeking to do as we wish to establish a Zion condition. He said: "We need to learn, practise, study, know, and understand how angels live with each other."[8] And Orson Hyde taught that if we will practice the way of life of the angels, they will indeed associate with and enlighten us: "The angels are not fond to descend to this world, because of the coldness of the spirit that reigns in it; they would rather remain in heaven around the throne of God, among the higher order of intelligences, where they can enjoy life, and peace, and the communion of the Holy One. [Nevertheless,] when they are sent, they will come. . . .

"These are some of the reasons [wars, contentions, coldness of

spirit, and so forth] why they do not mingle with us, why we cannot
see them. But let me tell you, brethren and sisters, if we will be
united as the heart of one man, and that general union of spirit, of
mind, be fastened upon the Lord Jesus Christ, we shall draw down
celestial intelligence by the Spirit of God, or by angels who surround
the throne of the Most High. It is an electric wire through which and
by which intelligence comes from heaven to mortals; it is only necessary for the word to be spoken, and the power of it is at once felt in
every heart."[9]

Consider these remarks from two prophets who had the vision of
Zion.

Joseph Smith: "By union of feeling we obtain power with God.
. . . Nothing is so much calculated to lead people to forsake sin as to
take them by the hand and watch over them with tenderness. When
persons manifest the least kindness and love to me, O what power it
has over my mind, while the opposite course has a tendency to harrow
up all the harsh feelings and depress the human mind. . . . It is the
doctrine of the devil to retard the human mind and retard our
progress by filling us with self righteousness. The nearer we get to
our heavenly Father the more are we disposed to look with compassion on perishing souls, to take them upon our shoulders and cast
their sins behind our back. . . . If you would have God have mercy
on you, have mercy on one another."[10]

Brigham Young: "If this people would live their religion, and
continue year after year to live their religion, it would not be many
years before we would see eye to eye; there would be no difference
of opinion, no difference of sentiment, and the veil that now hangs
over our minds would become so thin that we should actually see and
discern things as they are. . . . It is our privilege, for you and me to
live, from this day, so that our consciences will be void of offence
towards God and man; it is in our power to do so, then why don't
we?"[11]

"How prone we are to get out of the way, to depart from the love,
enjoyment, peace, and light that the Spirit of the Lord and of our
religion gives unto us. We should live so as to possess that Spirit

daily, hourly, and every moment. That is a blessing to us, which makes the path of life easy."[12]

"We will so improve, that, when a man rises here to pray, there will not be a desire from the heart of a man or woman but what is uttered by the one who is mouth. When we come to understanding, there will not be as many desires and prayers as there are people, while one is officiating as mouth for the whole; but when he who is mouth prays, every heart will wait until he utters a sentence, and that embodies what they also desire. When the sisters meet together and appoint one of their number to pray, they will never let a desire escape from the heart until they know what the mouth is praying for. Then they all will desire the same and pray for the same. This people are hastening to that degree of perfection."[13]

President Howard W. Hunter described God's way of dealing with us and thereby describes by implication how we could deal with one another to bring angels and miracles into our relationships: "God's chief way of acting is by persuasion and patience and long-suffering, not by coercion and stark confrontation. He acts by gentle solicitation and by sweet enticement. He always acts with unfailing respect for the freedom and independence that we possess. He wants to help us and pleads for the chance to assist us, but he will not do so in violation of our agency. . . .

"To countermand and ultimately forbid our choices was Satan's way, not God's, and the Father of us all simply never will do that. He will, however, stand by us forever to help us see the right path, find the right choice, respond to the true voice, and feel the influence of his undeniable Spirit. His gentle, peaceful, powerful persuasion to do right and find joy will be with us 'so long as time shall last, or the earth shall stand, or there shall be one man upon the face thereof to be saved' (Moro. 7:36)."[14]

When we live in patience and love with each other, in peace, meshing with those around us, not resisting them but supporting them, forgiving each other, speaking the words that evoke the Spirit, calling forth the divinity that lies in every person we know—no matter what his weaknesses—we live the spirit of at-one-ment with each other. The more we make each relationship sweeter and more tender and dear, the more we live at-one-ment. The more we lay down pride and old checklists of hurts and grievances, the more we send out healing, the more our relationships heal. Surely our personal connection with the Lord depends on our efforts toward resonance with each other.

If we absolutely knew that the Lord would send his Spirit any time that we began with thoughts, words, or actions, to generate the spirit of at-one-ment, why would we ever choose to generate something else? The Lord says, "Strengthen your brethren in all your conversation, in all your prayers, in all your exhortations, and in all your doings" (D&C 108:7).

What we find when we live the spirit of the at-one-ment is that it is a very empowering way of life. This mystery is not captured by words alone; it is experienced in spirit. We children of God already exist in a oneness with each other and with the Lord that is not apparent to the five human senses. The great majority of people in the world and even many members of the Church do not sense this oneness and live separated emotionally and spiritually from each other in spiritual death.

"I say unto you, were it not for the transgressions of my people, speaking concerning the church and not individuals, they might have been redeemed even now.

"But behold, they have not learned to be obedient to the things which I required at their hands, but are *full of all manner of evil,* and *do not impart of their substance, as becometh saints,* to the poor and afflicted among them; and are *not united* according to the *union* required by the law of the celestial kingdom; and *Zion cannot be built up unless it is by the principles of the law of the celestial kingdom;* otherwise I cannot receive her unto myself.

"And my people must needs be chastened until they learn obedience, if it must needs be, by the things which they suffer" (D&C 105:2–6).

But the enlightened person we are speaking of has been chastened enough. He's tired of spiritual death. He finds that when he sends out with love that which empowers others, he himself feels empowered and connected to the Lord Jesus Christ. The principles work. Elder Neal A. Maxwell wrote about the reality of this connectedness: "There may even be, more than we now know, some literalness in His assertion, 'Inasmuch as ye have done it unto one of the least of these my brethren, ye have done it unto me' (Matt. 25:40). We lack deep understanding of the implications of that remark of Jesus. As with so many things, He is telling us more than we are now prepared to receive."[15]

The person who has responded to the spirit of at-one-ment knows that he is living in spirit and participating increasingly in miracles. He has discovered a great secret that makes all the former pettiness and ego-striving seem far beneath his now higher purposes in life.

The Prophet Joseph taught the importance of being governed by higher purposes: "How vain and trifling have been our spirits, our conferences, our councils, our meetings, our private as well as public conversations—too low, too mean, too vulgar, too condescending for the dignified characters of the called and chosen of God, according to the purposes of His will, from before the foundation of the world!"[16]

One day, when we seek, ask, and knock, and the heavenly gate is opened, and we ask permission to enter, I think we will have to present something in ourselves recognizably heavenly in order to gain entrance. Let us practice at-one-ment here so that we will know how to act should we be admitted to heaven.

Notes

From *Spiritual Lightening* (Salt Lake City: Bookcraft, 1996), 75–91.

1. Brigham Young, Journal History, 23 February 1847, Salt Lake City, Church Historical Department, 2.
2. Parley P. Pratt, *Key to the Science of Theology* (Salt Lake City: Deseret Book Company, 1978), 31.
3. The emphasis in this verse and in subsequent scriptural passages in this chapter has been added by the author.
4. Brigham Young, in *Journal of Discourses*, 3:194.
5. *Teachings of the Prophet Joseph Smith*, 103; emphasis added.
6. *Teachings of the Prophet Joseph Smith*, 91; emphasis added.

7. David Whitmer, in B. H. Roberts, *Comprehensive History of the Church,* 6 vols. (Salt Lake City: Deseret News Press, 1930), 1:131.
8. Brigham Young, in *Journal of Discourses,* 6:76.
9. Orson Hyde, in *Journal of Discourses,* 1:126.
10. *Words of Joseph Smith,* 123.
11. Brigham Young, in *Journal of Discourses,* 3:194.
12. Ibid., 7:238.
13. Ibid., 6:76–77.
14. Howard W. Hunter, *Ensign,* August 1994, back page.
15. Neal A. Maxwell, "Becoming a Disciple," *Ensign,* June 1996, 12–13.
16. *Teachings of the Prophet Joseph Smith,* 137.

Chapter 15

SPIRITUAL CONTRADICTIONS AND THE QUEST FOR THE LORD JESUS

For many of us things haven't turned out the way we hoped or thought they would—or even the way we thought God promised they would. Many of our lives have not followed what we would have seen as an *ideal* pattern, and yet around us, others seem to have achieved ideals.

As things happen to us that disorient us spiritually, we do not feel "spiritual"—especially if we think we've created the trouble ourselves, or if we're angry at someone else who has made trouble for us, or if we do not understand what the Lord is doing. We often do not see or feel a spiritual purpose in what is going on. What shows up in our lives seems somehow contradictory to reason or expectation. Here are some things we might feel in the midst of spiritual contradictions:

That the Lord has breached our trust in him. We may be tempted to abandon our faith or refuse to serve or cooperate with this God—to want to keep our distance from him. We may even try to manipulate him by being angry and petulant and uncooperative.

That our trials are a sign that we are being punished or are somehow inferior spiritually or unworthy of the presence of the Lord, that there is something intrinsically wrong with us, or perhaps that spiritual experience is for other people, not for us.

That we have ruined something beyond repair, that we threw

something precious away by our own foolishness or rebellion—and that it is irretrievable.

And now with these perceptions it can be difficult to trust the Lord, because we seem to live in a world where anything can happen. How are we to understand this God?

What is the point of "seek[ing] this Jesus" (Ether 12:41)?

My purpose here is to tell you why we should seek this Jesus and why we can trust him, even when he doesn't do what we want him to. Nephi's brothers murmured "because they knew not the dealings of that God who had created them" (1 Nephi 2:12). We can forgive the Lord and trust him by learning more about his dealings.

Elder Richard G. Scott taught that adversity is of two different kinds: that which comes as consequence of our disobedience and that which comes because the Lord sees that we are ready to grow.[1] In a way, these two reasons for adversity can overlap. Both are trying to teach us spiritual principles that we may have ignored or do not know. As long as we are human beings, we will make mistakes that come from either ignorance of spiritual law or ignoring it. I'm speaking not of the obvious "big" sins, but of those that we barely acknowledge as sin, or may not acknowledge at all. To a much greater degree than we may have recognized, the universe is dedicated to bringing to our attention, *through our experiences,* that which we have not yet faced or dealt with—even things we might consider inconsequential. For the Lord to get our attention and for us to feel the need to change, these experiences must have impact. All this for one main purpose. Jesus teaches John the principle on the edge of the Jordan as John protests baptizing him: "It becometh us to fulfil all righteousness" (Matthew 3:15). A divine universe prods us to recognize and fulfill all righteousness so that, as at the Savior's baptism, the powers of heaven can be manifested.

Our experiences are designed to teach us to go the distance, to come up to the full measure of our godly creation, to "fulfil all righteousness"—again, so that the powers of heaven can be manifested. Jesus went the whole distance; for us there is a distance yet to go.

Our spirit knows when we're not going the distance in a

relationship, a responsibility, a trust, or a commandment. This knowledge may be buried in the unconscious, but one's spirit, connected to the Father, never forgets. This unconscious knowing about our failures clouds our spiritual processes and constitutes the mists of darkness. A benevolent universe conspires to bring to pass the experiences which will jolt our conscious mind into awareness. Why? It wants to show us what is not working. Without conscious awareness of the patterns that are *not* working in our life, we cannot change. But concious awareness can be followed by a better choice. And what if we don't make a better choice? When we're behaving badly, we feel the Spirit say to us, "You could change this behavior right now and avert the painful consequences which must come." Not to make a better choice means that sooner or later, one way or another, we will be kicked very hard from behind. Therefore, we might say that we make choices and then the divine universe responds (1 Nephi 2:20; Mosiah 7:29; Alma 41:5).

Our Father, loving us with a perfect love, will do whatever is necessary to bring us to awareness, change, growth, and the capacity for joy. One of the most important functions of life's experiences is that they reveal ourselves to ourselves—if we are open to the truth.

Elder Charles Penrose taught: "Here in the darkness, in the sorrow, in the trial, in the pain, in the adversity, we have to learn what is right and distinguish it from what is wrong, and lay hold of right and truth and learn to live it. . . . If we have any evil propensities . . . we have to grapple with them and overcome them. Each individual must find out his own nature, and what there is in it that is wrong, and bring it into subjection to the will and righteousness of God."[2]

Similarly, Elder George Q. Cannon wrote: "Whatever fate may threaten us, there is but one course for men of God to take, that is, to keep inviolate the holy covenants they have made in the presence of God and angels. For the remainder, whether it be life or death, freedom or imprisonment, prosperity or adversity, we must trust in God. We may say, however, if any man or woman expects to enter into the Celestial Kingdom of our God without making sacrifices and without being tested to the very utmost, they have not understood the

Gospel. If there is a weak spot in our nature, or if there is a fiber that can be made to quiver or to shrink, we may rest assured that it will be tested. Our own weaknesses will be brought fully to light, and in seeking for help the strength of our God will also be made manifest to us."[3]

Elder Cannon also taught: "Every Latter-day Saint who gains a celestial glory will be tried to the very uttermost. If there is a point in our character that is weak and tender, you may depend upon it that the Lord will reach after that, and we will be tried at that spot for the Lord will test us to the utmost before we can get through and receive that glory and exaltation which He has in store for us as a people."[4]

Each experience has a hidden question: "Will you accept this new revelation about yourself and about Me and my purposes?" Then we might reply with our spiritual knowing, "Thank thee, Father, for being honest with me."

"Nevertheless the Lord seeth fit to chasten his people; yea, he trieth their patience and their faith. Nevertheless—whosoever putteth his trust in him the same shall be lifted up at the last day" (Mosiah 23:21–22).

This preceding passage suggests that not every trial represents the consequence of past neglect. The Lord may be strengthening faith or stretching the capacity of a person to sacrifice for someone else's benefit. For example, we see in the scriptures accounts of trials that the righteous are subject to in order to bless or teach someone else. Nephi is bound so that his brothers can learn about consequences (1 Nephi 18:11)—"nevertheless, the Lord did suffer it"—that is to say, Nephi could not have been bound, because of his righteousness, except the Lord would allow it. See also Alma 24:24–27, in which many Anti-Nephi-Lehies lose their lives, yet many more Lamanites are converted by their nonresistance. Another scriptural example is Acts 16:23–34, where the Apostle Paul submits to beatings and imprisonment so that the jailer and his family can receive baptism.

But each trial is doubtless designed to expand awareness and sensitivity to spiritual things, and to carry us with power toward greater godliness. Adversity can also prepare one to hear the word of God

(Alma 32:6) and to overcome "dull[ness] of hearing" (Hebrews 5:11). As Elder Neal A. Maxwell says, "Given all you and I yet lack in our spiritual symmetry and character formation, no wonder God must use so intensively the little time available to develop each of us in this brief second estate. One's life, therefore, is brevity compared to eternity—like being dropped off by a parent for a day at school. But what a day!

"For the serious disciple, the resulting urgency means there can be few extended reveries and recesses and certainly no sabbaticals—all this in order to hasten God's relentless remodeling of each of us. Reveries and special moments may come, but they are not extended. Soon the drumroll of events, even difficulties, resumes. There is so much to get done in the brief time we have in this mortal classroom. . . . God is very serious about the joy of His children."[5]

Two things are true at once with respect to God's relationship to his children.

The first truth is based on these lines from scripture: "I am the true light that is in you, and . . . you are in me; otherwise ye could not abound" (D&C 88:50; see also vv. 41, 49). We can accept that Jesus is somehow enmeshed in our being, is conscious of each of us in a very individual way, and is trying to draw us into consciousness of him, into conscious experience with him (D&C 88:63–69).

The second truth: This loving God requires some tough challenges of us in order to make us into what we were foreordained to be, because "all intelligent beings who are crowned with crowns of glory, immortality, and eternal lives must pass through every ordeal appointed for intelligent beings to pass through, to gain their glory and exaltation."[6] Apparently every being who has attained to godhood had to pass first through trials in a fallen world. Indeed, Brigham taught that if there were any other way of bringing us to exaltation, except through trials in a fallen world, the Lord would have done it: "If man could have been made perfect, in his double capacity of body and spirit, without passing through the ordeals of mortality, there would have been no necessity of our coming into this state of trial and suffering. Could the Lord have glorified his children in spirit, without a body like his own, he no doubt would have done so."[7] In short, "Every trial and experience you have passed through is necessary for your salvation."[8]

The Brethren have also spoken about Abrahamic trials, which some of us may be honored to experience. John Taylor quoted the Prophet Joseph: "'You will have all kinds of trials to pass through. And it is quite as necessary for you be tried as it was for Abraham and other men of God, and . . . God will feel after you, and He will take hold of you and wrench your very heart strings, and if you cannot stand it you will not be fit for an inheritance in the Celestial Kingdom of God.'"[9]

In another place Elder Taylor said, "I speak of these things to show how men are to be tried. I heard Joseph Smith say—and I presume Brother Snow heard him also—in preaching to the Twelve in Nauvoo, that the Lord would get hold of their heart strings and wrench them, and that they would have to be tried as Abraham was tried. Well, some of the Twelve could not stand it. They faltered and fell by the way. It was not everybody that could stand what Abraham stood. And Joseph said that if God had known any other way whereby he could have touched Abraham's feelings more acutely and more keenly he would have done so. . . . We cannot conceive of anything that could be more trying and more perplexing than the position in which he was placed."[10]

Often in the midst of trials we'd just appreciate some clarity. Sometimes clarity is not offered; the veil is left in place. "The knowledge of our former state has fled from us. . . . [T]he veil is drawn between us and our former habitation. This is for our trial. If we could see the things of eternity, and comprehend ourselves as we are; if we could penetrate the mists and clouds that shut out eternal realities from our gaze, the fleeting things of time would be no trial to us, and one of the great objects of our earthly probation or testing would be lost. But the past has gone from our memory, the future is shut out from our vision and we are living here in time, to learn little by little, line upon line, precept upon precept."[11]

Elder Maxwell explained, "If everything in one's immediate

context were constantly clear, God's plan would not work. Hard choices as well as passing through periodic mists of darkness are needed in order to maintain life's basic reality—that we are to overcome by faith."[12]

"[As Brigham Young] said even in the midst of deep and discouraging blackness we are to trust in the Lord in order to show that we are 'a friend of God' by being 'righteous in the dark.' On another occasion, Brigham Young called for us to be faithful even if circumstances are 'darker than 10,000 midnights.'"[13] Someone asked President Young, "Why are [we] left alone and often sad?" Brigham Young replied, "To practice [a person] to be righteous [even] in the dark."[14]

Therefore, while he is an all-loving and merciful God, drawing us close, our Heavenly Father also is dedicated to our godly development and has a perfect understanding of how that is to be accomplished in each individual case. So he shapes our individual growing and sanctifying experiences to include light and dark, clarity and contradiction.

And then—and this is the most important part—in the midst of those orchestrated trials, *he strives with us* and is striving with us to see us through this often painful, refining process. His purpose is not just to help us to survive but to help us to keep covenant and to fulfill all righteousness.

We are at times required to trust without understanding. Except this much we can understand: whatever has come upon a person is not random. The Lord has allowed it and will give us grace to bear it and benefit from it for our advancement. "And when He is calling us to pass through that which we call afflictions, trials, temptations, and difficulties, did we possess the light of the Spirit, *we would consider this the greatest blessing that could be bestowed upon us.*"[15]

We can begin to see that we could even look at our painful experiences, as well as the people who participated in them, and not only thank the Lord for being honest with us, but thank the other person for being our friend, for serving us, for helping us to see a truth we needed.

I have heard people say, "The Lord has brought this on me to humble me." Perhaps that is so, but often what the person seems to mean is that the Lord has brought on this trial to smash him, to defeat him, to squash him like a worthless worm. But this I know: Our trials, whatever their source, have a spiritually strengthening purpose only and are the evidence that we are worth preparing for exaltation.

Still, however beneficial all this experience, the Lord acknowledges that the chastening and refining can hurt. One of the themes of the Isaiah portions of the Book of Mormon deals with Israel's feelings of having been forsaken by God. He poignantly replies to the cry of pain:

"But, behold, Zion hath said: The Lord hath forsaken me, and my Lord hath forgotten me—but he will show that he hath not.

"For can a woman forget her sucking child, that she should not have compassion on the son of her womb? Yea, they may forget, yet will I not forget thee, O house of Israel.

"Behold, I have graven thee upon the palms of my hands; thy walls are continually before me" (1 Nephi 21:14–16; see also 1 Nephi 20:10).

Here we feel that the Lord must allow us to pass through pain, just as he himself did, in order to achieve our lofty objective. Nevertheless the Lord does all things for men *"which were expedient for man to receive"* (1 Nephi 17:30).[16] He cannot remove it all without interfering with the purposes of the mortal probation. Even so, as the Book of Mormon teaches, "And if it so be that the children of men keep the commandments of God he doth nourish them, and strengthen them, and provide means whereby they can accomplish the thing which he has commanded them; wherefore, he did provide means for us while we did sojourn in the wilderness" (1 Nephi 17:3).

He strives with us and for us, even when we don't perceive it.

Much of what happens in this world can only begin to be understood in the context of arrangements made in the premortal world. There each of us accepted a work that entailed certain good each of us would do here—much of it in the face of opposition. A point of doctrine: if the Lord were to do all that good *for* us, by the eternal law of restoration (i.e., Alma 41:15, "For that which you do send out shall return unto you again, and be restored"), good would be restored to *Him*. He is trying to help each of us make and keep covenants and, in spite of earth and hell, to do our own premortally covenanted good, so that good can be restored to us. The Lord is trying to help us to accomplish what we covenanted to do in the life before because our exaltation rests on its accomplishment.

We can make our trials much worse than they have to be. Paying attention to the dark side can make hard trials harder or create trials where there don't have to be. We remember that in the premortal world, the battles of good and evil were fought, not with weapons as we know them, but with mental warfare: powerful intelligences tried to recruit less experienced intelligences to the dark side. In the mortal probation, the same battle continues. It still consists of mental and emotional warfare from seen and especially unseen representatives of the dark side. Satan is enticing each of us to consider *his* things— his temptations, worries—trying to demoralize us; and he succeeds as we consider his voice and his thoughts so that we waste a part of our energies and reserves. Satan's thoughts are made very compelling even though many of them are thoughts of misery or lead to misery. But here is one of life's major tutorials of eternal significance: we have the power to *turn him off*. We learn to say, "No, I will not consider that thought." We substitute higher thoughts and prayers. Satan will attempt to drive upon us and he will rage and he will not accept this. He will come again and again. Satan comes in on us uninvited with a thousand capabilities. We must use our agency; God will not intercede unless we invite the Spirit. When we learn to keep the mind occupied with God's things, gradually the Spirit comes and takes up residence, bringing comfort and direction and fulfilling ancient promises.

We must understand that there are only two spirits abroad in the world (Alma 5:39). We do not live at any time in a vacuum. While the Lord's Spirit is striving with us to help us do our good, the spirit of the adversary is seeking constant access to the mind, heart, and spirit. If we are not consciously under the influence of the Lord Jesus

Christ, we are taking our signals from the adversary and paying a heavy price without knowing why.

If we are not deliberately allowing Christ to lead us, we are driven by our own appetites and by the adversary. It is a matter of what one is tuning to, to what one allows time and space in one's mind. Satan would have us believe that we are victims of our own emotions, and that these emotions represent reality. But none of us is a victim—we were given dominion over Satan and over his thought-world; many of our emotions reflect only the miserable influence, delusions, and programming of the adversary. One's mental world is very powerful—maybe all powerful.

Here is an important question: *What am I immersed in mentally—the world spirit, or the Christ spirit?* We can take in large doses of the world through its various media if we choose, but we will pay in a diminished amount of faith and diminished control over the quality of our emotional and spiritual lives. The mind of the world is continually seeking access to our mind. It is aggressive and sometimes subtle. So one needs to take care not to feed one's mind with the media: television, movies, videos, magazines that speak for the telestial world. These will fill the mind with the spirit of the world and neutralize our faith—they are the agents of the adversary. I'm not speaking only of that which is blatantly ugly and obscene but also of that which is spiritually shallow—which means that there is really not much that the world spirit has to offer us. If we want effective faith, we must learn to guard the mind carefully and be very picky—to allow in—insofar as possible—only that which is Christlike in all its wonderful possibilities for nobility, beauty, humility, and compassion.

Another thing that the adversary seeks to get us to do is to obsess over problems and deprivations; this mental path is made very compelling. Many of us are drawn to negative thoughts and memories unless we have practiced otherwise. But really, obsessing, worrying, stewing over things won't change them; going over and over them in our minds finally deepens the mists of darkness that blind the eyes and harden the hearts of men.

We are free to choose what we will think about, but it is very helpful—maybe essential—to learn thought substitution and to practice flooding the mind with urgent prayer so as to block the fiery dart of the adversary. We can sense that the adversary's missile is approaching before it penetrates its target (one's spirit). We must quench it before the assault affects our physical and spiritual chemistry, setting up hormonal and other emotional processes that can be so hard to recover from. "Father, help me, help me!" This flooding with urgent prayer puts up the shield of faith, deflects the fiery dart, and protects the follower of Christ (D&C 27:17).

And we can have thoughts ready to substitute. We all have things treasured up that bring up our spiritual energy and remind us of who is really in control in the universe. We don't have to give our power to Satan.

It's important to know that we could not protect ourselves against Satan if it were not for the priesthood and the Atonement. Satan would have swallowed us whole, and we would even now be angels to a devil were it not for the power of the Atonement, which makes it possible to protect ourselves from being recruited by Satan. One of the purposes of the mortal probation is to identify and reject him in all his manifestations. The adversary is always seeking access to us, trying to bring us down; "he seeketh that all men might be miserable" (2 Nephi 2:27).

With a little effort, each one of us can identify our own personal demons. Which thoughts take me on that downward path? Those are my demons, and we are not talking figuratively. Jesus comes to "cast out devils, or the evil spirits which *dwell* in the hearts of . . . men" (Mosiah 3:6). Everyone is more or less under the adversary's influence until we make a conscious decision to escape.

We must feed the mind continually with God's things or we will not endure the testing of the mortal probation. We *can* transcend the natural-man thought world with careful selection of thoughts. We know how to identify the enemy's thought world: spiritual carelessness, untruths, hatred in all its shades, despair, fear, lust, greed, willfulness, selfishness, careless behavior with others' feelings, being critical and prideful, and enmity in all its forms. With the help of the Spirit of the Lord, we can live mentally and spiritually in a dimension apart from the spirit of the world—or we will be sucked into its negative emotional states.

The scriptures command us to be happy as they teach us what to think about:

"Look unto me in every thought; doubt not, fear not" (D&C 6:36).

"[C]heer up your hearts, and remember that ye are free to act for yourselves" (2 Nephi 10:23).

"Let your hearts rejoice" (2 Nephi 9:52).

Our happiness is our own responsibility. The power comes with the command (1 Nephi 3:7). Everything has been provided to make it possible. The decision to be happy and trust the Lord opens one's spirit to the counsel and comfort and the presence of the Lord.

Reality and life are fluid and powerfully influenced by our choices, ever reconfiguring themselves. As we change inwardly, our circumstances change outwardly. When we begin to give ourselves to full discipleship, our circumstances do change. Then the focus of our lives becomes the quest of the Lord Jesus Christ. The more we are consumed in Christ (to use President Ezra Taft Benson's phrase)[17] the more we see that things are working together for our specific blessing and that setbacks are only apparent—that there is a benevolent purpose. We can live knowing, through the details of our life, that our life is in Christ's benevolent hands. This blessed condition we can elect to initiate at any time: "Fear not, Daniel: for from the first day that thou didst set thine heart to understand, and to chasten thyself before thy God, thy words were heard, and I am come for thy words" (Daniel 10:12).

Unless we are on the path of honest, deliberate, determined discipleship, willing to sacrifice and forgo the deadliness of the world to find Christ personally, willing to deal with the lessons with which the Lord is confronting us, willing to align ourselves with the higher purposes of our existence, we will not have the fulness we want— we will languish on the periphery. These are the magic words:

"Father, I will serve thee, no matter what." The pursuit of disciple-ship to the Lord Jesus Christ—and all that that entails—is finally the only thing that gives life meaning.

Brigham Young describes this comprehensive consecration: "I will tell you one mark you have got to come to, in order to do right. If you can bring yourselves, in your affections, your feelings, your passions, your desires, and all that you have in your organization, to submit to the hand of the Lord, to his providences, and acknowledge his hand in all things, and always be willing that he should dictate, though it should take your houses, your property, your wives and children, your parents, your lives, or anything else you have upon the earth, then you will be exactly right; and until you come to that point, you cannot be entirely right. That is what we have to come to; we have to learn to submit ourselves to the Lord with all our hearts, with all our affections, wishes, desires, passions, and let him reign and rule over us and within us, *the God of every motion:* then he will [and can] lead us to victory and glory; otherwise he will [and can] not."[18]

I know that Jesus Christ lives, that he is merciful, and that he is conscious without interruption of each of us—"he remembereth every creature of his creating" (Mosiah 27:30). He invites us to become conscious of him.

Finally, consider all those gathered at the throne of God; they will be those like you and me, ordinary people, who have tried to endure the crosses of the world as well as powerful temptations, perhaps the trials of singleness or the pain of failed relationships. They will be those who may have sinned serious sin but who have been led by the Savior to repent and be faithful, to develop in godly ways, and to do their good in the midst of encircling darkness. There won't be one person there who hasn't suffered contradictory trials, a profound sense of inadequacy, and periodic mists of darkness. It will have been the Savior, with his incomparable grace, often unac-knowledged by the very one he was helping, who pulled each one through the tutorials of the mortal probation, making it possible to be fitted for the greater things of the world to come.

I know that we can have experiences with the Lord which lift the

veil a little, and that when it is lifted, we understand that he understands and knows it all, and has been striving and participating with
us all along. "And now, I would commend you to seek this Jesus of
whom the prophets and apostles have written, that the grace of God
the Father, and also the Lord Jesus Christ, and the Holy Ghost, . . .
may be and abide in you forever. Amen" (Ether 12:41).

Notes

This is a previously unpublished paper.

1. Richard G. Scott, in Conference Report, October 1995, 18–21.
2. Charles W. Penrose, in *Journal of Discourses*, 26:28.
3. George Q. Cannon, *Gospel Truth: Discourses and Writings of President George Q. Cannon*, ed. Jerreld L. Newquist, 2 volumes in 1 (Salt Lake City: Deseret Book Co., 1987), 304.
4. Cannon, *Gospel Truth*, 81.
5. Neal A. Maxwell, "Becoming a Disciple," *Ensign*, June 1996, 16, 17.
6. Brigham Young, in *Journal of Discourses*, 8:150.
7. Ibid., 11:43.
8. Ibid., 8:150.
9. Joseph Smith, as quoted by John Taylor, in *Journal of Discourses*, 24:197.
10. John Taylor, in *Journal of Discourses*, 24:264.
11. Charles W. Penrose, in *Journal of Discourses*, 26:28.
12. Neal A. Maxwell, *Lord, Increase Our Faith* (Salt Lake City: Bookcraft, 1994), 110.
13. Brigham Young, in *Journal of Discourses*, 3:207, as quoted in Neal A. Maxwell, *That Ye May Believe* (Salt Lake City: Bookcraft, 1992), 3.
14. Brigham Young, as quoted in Maxwell, *That Ye May Believe*, 194–95.
15. Brigham Young, in *Journal of Discourses*, 2:303; emphasis added.
16. The emphasis in this verse and in subsequent scriptural passages in this chapter has been added by the author.
17. Ezra Taft Benson, "Born of God," *Ensign*, July 1989, 4.
18. Brigham Young, in *Journal of Discourses*, 5:351–52; emphasis added.

Chapter 16

SPIRITUAL TWILIGHT AND THE CONTEST WITH EVIL

We know before we come to the earth what we want to accomplish here; we come with a definite purpose. But when we get here, we can't remember it. Then the darkness moves in, taking advantage of our having forgotten who we are and why we're here. This darkness, combined with the loss of memory, produces a sort of spiritual sleep in the fallen man which can keep him in what might be termed a spiritual twilight—not full darkness, but neither that resplendent life in Christ. A person can get stuck in this half-light.

The gospel of Jesus Christ is the way out, but for many Latter-day Saints this twilight can be a prolonged experience in which one is stalled—recognizing one's spiritual isolation but not yet reaching to the solution. This is the state where the soul cries out, "Is this all there is?" We get stuck here because we are doing some things right, we are going through some motions, we are feeling occasional Spirit, we seem to be on the path; but still, there's that nagging discontent. The empty "soul hath appetite" (2 Nephi 27:3).

However, a person can and must get beyond this semi-darkness if he wants the life that God is offering him. This escape takes knowledge and effort and divine help. It is, after all, our own ignorance or disobedience that keeps us in the twilight, but in addition there are unseen forces that are bent on keeping us there. Having forgotten all, man arrives on earth with a strong inclination to pride, so his mind is a fertile field for the adversary's temptations; the harvest is sin and captivity (1 Nephi 12:19). These unclean influences are very subtle; the adversary blends unwholesome thoughts and feelings into those of our own spirits. While we go along not even knowing we are under attack, aggressive emissaries of evil are pressing us always toward the same end: self-destruction through our own disobedience. But sometimes just a new insight can bring a transformation, reduce pride, fortify the willing, and clarify and smooth the path to fulness. Spiritual discernment and resolve can deliver us from the spiritual twilight. I'm going to focus here on knowledge that we must have to transcend darkness.

Our spiritual discernment is not developed when we are born. Like babes lying in their cradles, we are immersed in stimuli we don't yet distinguish. Spiritual influences from the cosmic ocean wash upon our souls, and in time, motivated by various needs, we learn to identify selected sights, sounds, and sensations in the material world. So also, maturing spiritually, we learn to perceive and discern that which lies just beyond our physical senses. The ability to discern is not a luxury, it is a necessity. One of the greatest mysteries to the fallen and natural man is that he is spiritually dead, which means that he is asleep to spiritual realities and is unable to participate consciously with the benevolent cosmic forces in which he is immersed. He struggles in spiritual bondage that he only barely recognizes and from which he alone cannot extricate himself (see, e.g., Alma 5:6–9). He is in bondage to his spiritual ignorance because forces of evil play upon him and he is helpless against them. One of the major purposes of this life is to teach us to discern between dark and light.

Young Joseph learned early to discern the light. When Moroni appeared the first time to Joseph, he gave him specific instructions and a warning "that Satan would try to tempt me (in consequence of the indigent circumstances of my father's family), to get the plates for the purpose of getting rich. This he forbade me, saying that I must have no other object in view in getting the plates but to glorify God, and must not be influenced by any other motive than that of building his kingdom; otherwise I could not get them" (Joseph Smith—History 1:46).

But on the morning of September 22, 1823, as Joseph walked the distance of two or three miles from his home to the place where the plates were buried in the Hill Cumorah, his young and inexperienced mind struggled between two alternatives. Oliver Cowdery reported, "The mind of man is easily turned if it is not held by the power of God through the prayer of faith and you will remember that I have said that two invisible powers were operating upon [Joseph's] mind during his walk from his residence to Cumorah, and that the one urging the certainty of wealth and ease in this life had so powerfully wrought upon him that the great object so carefully and impressively named by the angel, had entirely gone from his recollection that only a fixed determination to obtain now urged him forward.

"After arriving at the repository, a little exertion in removing the soil from the edges of the top of the box, and a light pry, brought to his natural vision its contents. No sooner did he behold this sacred treasure than his hopes were renewed, and he supposed his success certain . . . without once thinking of the solemn instruction of the heavenly messenger, that all must be done with an express view of glorifying God.

"On attempting to take possession of the record a shock was produced upon his system, by an invisible power, which deprived him, in a measure, of his natural strength. He desisted for an instant, and then made another attempt, but was more sensibly shocked than before. . . . He . . . made the third attempt with an increased exertion, when his strength failed him more than at either of the former times, and without premeditating he exclaimed, 'Why can I not obtain this book?' 'Because you have not kept the commandments of the Lord,' answered a voice, within a seeming short distance. He looked, and to his astonishment, there stood the angel who had previously given him the directions concerning this matter. In an instant, all the former instructions . . . were brought to his mind. . . . At that instant he looked to the Lord in prayer, and as he prayed darkness began to disperse from his mind and his soul was lit up as it was the evening before, and he was filled with the Holy Spirit; and again did the Lord manifest his condescension and mercy: the heavens were opened and the glory of the Lord shone round about and rested upon him. While he thus stood gazing, and admiring, the angel said, 'Look!' and as he thus spake he beheld the prince of darkness, surrounded by his innumerable train of associates. All this passed before him, and the heavenly messenger said, 'All this is shown, the good and the evil, the holy and impure, the glory of God and the power of darkness, that you may know hereafter the two powers and never be influenced or overcome by that wicked one. . . . You may learn from henceforth that his ways are to destruction, but the way of holiness is peace and rest. You now see why you could not obtain this record. . . .

"'You have now beheld the power of God manifested and the power of Satan: you see that there is nothing that is desirable in the works of darkness; that they cannot bring happiness; that those who are overcome therewith are miserable, while on the other hand the righteous are blessed with a place in the kingdom of God where joy unspeakable surrounds them.' . . .

"God knowing that Satan would thus lead [Joseph's] mind astray, began at the early hour, that when the full time should arrive, he might have a servant prepared to fulfill his purpose. So, however affecting to his feelings this repulse might have been . . . he had learned by experience, how to discern between the spirit of Christ and the spirit of the devil."[1]

The Lord had to teach Joseph promptly and dramatically; even so, his education in discernment undoubtedly continued—as it continues for us.

The need to discern started in the life before. The Prophet Joseph taught, "Before the foundation of the Earth in the Grand Counsel . . . the Spirits of all Men were subject to oppression & the express purpose of God in Giveing [them] a tabernicle was to arm [them] against the power of Darkness."[2] Joseph taught that all God's acts are for the benefit of inferior intelligences; and when he saw that those intelligences had not power to defend themselves against those that had a tabernacle, he called them together in council and agreed to form them tabernacles. "For it is a Natureal thing with those spirits that has the most power to bore down on those of Lesser power."[3] This bearing down of more powerful spirits on less experienced ones is the nature of the war that took place in heaven—one more powerful mind trying to recruit a less mature spirit. The weapons have not changed.

One of the major purposes, then, in sending us to earth was to give us knowledge through experience with evil and to arm us against it, now and forever. Even spirits who were valiant in the premortal world slip into forgetfulness in their descent to the earth and find themselves entangled in evil. The younger Alma, a very wicked man in his early days, described in retrospect the bitterness of the days of his spiritual unconsciousness: "I was in the darkest abyss; but now I behold the marvelous light of God. My soul was racked with eternal torment; but I am snatched, and my soul is pained no more" (Mosiah 27:29). He learned that the Redeemer "remembereth every creature of his creating" (Mosiah 27:30).

That is, even though the earthly sleep is part of the divine design, even though it was necessary for man to condescend to taste the bitterness of sin, sooner or later the Father reaches after each of his children to try their desire. President Joseph F. Smith wrote, "After descending below all things, Christ-like, we might ascend above all things, and become like our Father, Mother and Elder Brother, Almighty and Eternal!"[4] Brigham Young exclaimed similarly, "I . . . praise God in the highest for His great wisdom and condescension in suffering the children of men to fall into the very sin into which they had fallen, for He did it that they, like Jesus, might descend below all things and then press forward and rise above all."[5]

Of course, our first parents rejoiced over this same opportunity: "Blessed be the name of God, for because of my transgression my eyes are opened, and in this life I shall have joy, and again in the flesh I shall see God," exclaimed Adam. And Eve said, "Were it not for our transgression we never should have . . . known good and evil, and the joy of our redemption" (Moses 5:10–11). They both mention joy. "You cannot give any persons their exaltation unless they know what evil is, what sin, sorrow, and misery are, for no person could comprehend, appreciate and enjoy an exaltation upon any other principle."[6]

No matter what may have been in our past, we are offered a victory over evil. The Prophet Joseph defined salvation with respect to our power over evil: "Salvation is nothing more or less than to triumph over all our enemies [evil] & put them under our feet & when we have power to put all enemies under our feet in this world & a knowledge to triumph over all evil spirits in the world to come then we are saved."[7]

Knowledge of the spirits is what we need: "If you wish to go whare God is you must be like God or possess the principles which God possesses for if we are not drawing towards God in principle we are going from him & drawing towards the devil. . . . A man is saved no faster than he gets knowledge for if he does not get knowledge he will be brought into captivity by some evil power in the other world as evil spirits will have more knowledge & consequently more power than many men who are on the earth."[8]

It is important for us to know that the spirits who followed Satan in the great rebellion in heaven got here even before Adam and Eve and are still here preying on you and me, seeking whatever degree of possession they can get. "The Devil with one-third part of the spirits of our Father's Kingdom got here before us, and we tarried there with our friends, until the time came for us to come to earth and take tabernacles, but those spirits that revolted were forbidden ever to have tabernacles of their own. You can now comprehend how it is that they are always trying to get possession of the bodies of human beings."[9] And we are their inescapable victims until we learn to discern what is *of God* from what is *not of God*. There are only two options and only two voices: "Behold, I say unto you, that the good shepherd doth call you; . . . and if ye will not hearken unto the voice of the good shepherd, . . . ye are not the sheep of the good shepherd. And now if ye are not the sheep of the good shepherd, of what fold are ye? Behold, I say unto you that the devil is your shepherd, and ye are of his fold" (Alma 5:38–39).

We have to know the enemy. He deals in anger, lies, and justification of sin. To keep us chained, he may rage in the heart to cause us to destroy relationships; or he may pacify us into carnal security so that we are numb to the real causes of our malaise. Leading us carefully, he may flatter us to soothe the smarting child of God lest the child discover his real problem and repent. And while we are in hell itself, he tells us that there is no hell (2 Nephi 28:20–22). He is indeed a subtle enemy (Moses 4:5). He seduces a person into disobedience. He says, "I am no devil, for there is none—and thus he whispereth in their ears, until he grasps them with his awful chains, from whence there is no deliverance" (2 Nephi 28:22). And the reason he is subtle is that he does not want to be detected as separate from the person he is seducing. If a person could detect him as a separate voice, he would know to dismiss him. We can learn to recognize his voice. Just this knowledge alone gives us considerable power. That is, if we are not consciously cultivating the Lord's shepherding, we may know for a surety that we are taking signals from the adversary and reaping the physical, emotional, and spiritual consequences.

The scriptures urge us, "Humble yourselves before the Lord, and call on his holy name, and watch and pray continually, that ye may not be tempted above that which ye can bear, and thus be led by the Holy Spirit, becoming humble, meek, submissive, patient, full of love and all long-suffering" (Alma 13:28).

We have to learn something about the adversary's objectives. When the Lord gave Adam dominion over the earth, one of Lucifer's first designs was to wrest this power from Adam and his posterity. Adam's dominion over every living creature was a function of his priesthood—in fact, the keys of the First Presidency over all the earth, from generation to generation, which keys he had obtained in the creation before the world was formed.[10] It is easy to see why Satan wants to wrest Adam's dominion, eradicate the true priesthood from the earth, and set up his own dominion. There are many kinds of evil intelligences throughout the cosmos, and they have their own power, but the priesthood of the Lord Jesus Christ has ascendancy over all spirits, as Satan well knows. The eternal priesthood which rests upon the faithful elders of the Church is above and presides over every other power.[11] And the name of Jesus Christ has power over all spirits. With our narrow vision here on the earth, we do not realize the grandeur and power we have been given in the true Church of Jesus Christ, but we can be so grateful for all covenants and ordinances that give us power over these spirits. They are dangerous to the ignorant and unarmed.

The only means by which the adversary expects to accomplish his revolution is to take possession of the tabernacles of Adam's posterity and thus exercise his dominion over the earth through those whom he has duped. Although Satan, because of the Lord Jesus Christ, will never fully succeed in this overthrow, he does nevertheless achieve significant lesser victories in order to enlarge his own dominion and empire in both time and eternity (2 Nephi 2:29). In many instances the hosts of evil have succeeded in gaining possession of the spirits and bodies of men.

Elder Erastus Snow taught, "They do not altogether get possession of the tabernacles of men, only in isolated cases. There are cases in which it seems that these spirits so far control the tabernacles of men as to find the natural spirit that owns these tabernacles and suspend the operations of their functions, and usurp the control of the functions of the body, and make these organs of speech speak the language of devils, and make these tabernacles perform the wicked works of the evil one; while the spirit that owns, and should control this tabernacle, is bound, as it were, hand and foot; and where these powers and functions are thus suspended . . . and they are under the dominion of devils."[12]

Then Elder Snow offered this wake-up observation: "But others, and this embraces all of us, are more or less influenced by evil spirits, that prompt and lead to passions, and the lusts of the flesh; and to do many things in violation of the true laws of life and health, and of peace and glory and exaltation, and these evils to which we are prompted through the influence of these spirits are designed, little by little, to bring us into bondage, to sin and death, and to him who has the power of death, which is the devil."[13]

Here we have the fundamental source and dynamics of the spiritual twilight. We can appreciate even more the Lord's kindness in giving us principles that protect us or deliver us from the assaults of our enemies. We can exclaim with Nephi, "Awake, my soul! No longer droop in sin. Rejoice, O my heart, and give place no more for the enemy of my soul" (2 Nephi 4:28; see also vv. 27, 29–35).

We are careless and unwary in our associations one with another, not understanding how much the protection of the Spirit depends on whether we treat each other spiritually. "The men and women, who desire to obtain seats in the celestial kingdom, will find that they must battle . . . every day."[14] Sometimes the veneer of our spirituality is too thin to resist the daily abrasions we experience with each other. We are reactionary, forgetful, and inclined to depart from love and peace and the light of the Spirit. "We should live so as to possess that Spirit daily, hourly, and every moment . . . which makes the path of life easy."[15] Every person who strives to purify his life is closely watched by fallen spirits, who are never idle. "They are watching every person who wishes to do right, and are continually prompting them to do wrong. This makes it necessary for us to be continually on our guard—makes this probation a continual warfare . . . a spiritual warfare."[16]

Because everyone is under attack, we must encourage each other; we must strengthen, inspire, and praise spiritual excellence in one another. How important it is to fortify the spirit of tender connectedness with each other! As Joseph taught, the policy of the adversary is to separate, to isolate, to break up at-one-ment, to destroy.[17] The very words we speak are spirit and carry a spirit which stirs that same spirit in the person who hears them. Every time we speak we generate spirit. How easy it is to create a Spirit-filled atmosphere of peace and love. How mindful we want to be to choose our words with care so as to produce continually a higher spiritual influence for ourselves and those around us, thereby avoiding that despair caused by evil influences. Brother Brigham taught:

"If you feel evil, keep it to yourselves until you overcome that evil principle. This is what I call resisting the devil, and he flees from me. . . . When you are influenced by the Spirit of holiness and purity, let your light shine; but if you are tried and tempted and buffeted by Satan, keep your thoughts to yourselves—keep your mouths closed; for speaking produces fruit, either of a good or evil character. . . . You frequently hear brethren and sisters say that they feel so tried and tempted, and have so many cares, and are so buffeted, that they must give vent to their feelings; and they yield to the temptation, and deal out their unpleasant sensations to their families and neighbors. Make up your minds thoroughly, once for all, that if we have trials, the Lord has suffered them to be brought upon us, and he will give us grace to bear them. . . . But if we have light or intelligence—that which will do good, we will impart it. . . . Let that be the determination of every individual, *for spirit begets spirit—likeness, likeness; feelings beget their likeness.* . . . If then we give vent to all our bad feelings and disagreeable sensations, *how quickly we beget the same in others,* and load each other down with our troubles, and become sunk in darkness and despair! . . . In all your social communications . . . let all the dark, discontented, murmuring, unhappy, miserable feelings—all the evil fruit of the mind, fall from the tree in silence and unnoticed; and so let it perish, without taking it up to present to your neighbors. But when you have joy and happiness, light and intelligence, truth and virtue, offer that fruit abundantly to your neighbors, and it will do them good, and so strengthen the hands of your fellow-beings."[18]

President Young's advice might seem to fly in the face of current theories about man's emotional well-being. But it may be that voices of the adversary have taken advantage of our spiritual ignorance and have slowly programmed us to think and feel in certain ways that create our emptiness and hunger. We need to know that by ourselves we cannot entirely undo that programming, but our own efforts at repossessing the agency given to us plus the freely offered power of Christ will exorcise our (sometimes) longtime, dark companions.

"And he shall cast out devils, or the evil spirits which dwell in the hearts of the children of men" (Mosiah 3:6).

One thing we surely need is a sturdy vigilance. We are bombarded by influences: visual, aural, unseen—all to provide sufficient scope for man's choosing; but at the same time we have the divine power to be independent of unwanted influences (D&C 93:30) once we know them for what they are. There are spirits in the atmosphere, filled with an evil disposition, who seek to influence us. Elder George Q. Cannon taught: "If our eyes were open to see the spirit world around us, we should feel differently on this subject than we do; we would not be so unguarded and careless, and so indifferent whether we had the spirit and power of God with us or not; but we would be continually watchful and prayerful to our heavenly Father for His Holy Spirit and His holy angels to be around about us to strengthen us to overcome every evil influence."[19] Elder Cannon said that the adversary has numerous agencies at his command and "if we could see with our spiritual senses as we now see with our natural senses, we should be greatly shocked at the sight of the influences that prompt us to disobey the counsels of God or the Spirit of the Lord in our hearts. . . . He who is imbued with the Spirit of God is sensibly aware when the evil power approaches; but he does not welcome it to his bosom; he resists it with all the might and strength God has given unto him, and he obtains power over it, and it no more troubles him; if it does, its influence is more weakened than previously."[20]

One of the most characteristic signs that a person is under the influence of the dark side is that his desire for righteousness has been weakened or flattened, even his desire for life itself. The reason is that he has ceased to involve himself in a great cause and has succumbed to the spirit that seeks dominion in the earth. This flatness may initially seem harmless, but it is one of the early stages of an attitude that leads to misery and even destruction. "It is the duty of every one to labor day by day to promote each other's happiness, and also to study the well-being of mankind. When we take a course opposite to this, we become uneasy, unhappy and discontented; we are not satisfied with anything that is around us. . . . It is the spirit of the world, or that spirit which controls the world, which causes people to feel in this way; and unless they drive it far from them it will lead them down to sorrow, misery and death."[21] Living to bless other people keeps the darkness at bay.

So long as we live in the flesh, none of us, of the posterity of Adam, can be entirely free from the power of the devil. This being continually fretted and bothered is for our blessing, for we have seen here that one of the great objects God has in view in sending us to the earth is to give us experience in these influences that we may overcome them so that when we pass beyond the veil we may comprehend them to a greater extent than had we not come.[22] Our struggles here are not only for our protection and godly development now but are an indispensable preparation for the world to come. When the faithful members of the Church go into the spirit world, they carry with them the same knowledge, power, and priesthood they received here. If one gets his victory over the adversary in this world, he will have control over those evil spirits in the spirit world. "When this portion of the school is out, the one in which we descend below all things and commence upon this earth to learn the first lessons for an eternal exaltation, if you have been a faithful scholar, and have overcome, if you have brought the flesh into subjection by the power of the Priesthood, if you have honored the body, when it crumbles to the earth and your spirit is freed from this home of clay, has the devil any power over it? Not one particle."[23]

The Holy Ghost is the protection the Lord has conferred on us— but we must receive it by qualifying ourselves through obedience to the Lord's instructions. "When a person does fit and qualify himself, spiritual messages, waiting to be revealed, come to him. Then, and only then, is spiritual knowledge quickened into living comprehension leading to activity. When there is such correspondence between an individual and the spiritual world, the real joy of life appears. Otherwise, something is missing from our daily desire. We live incompletely. . . .

"Is it difficult to . . . qualify oneself spiritually? Nothing is easier or more enjoyable. When there is harmony between the instrument and the pounding message [of the Spirit], there is joy in the heart. The world's confusion roots in discord, lack of harmony [with that Spirit]. To be out of focus or to live in the midst of static is to be in semi-darkness and chaos."[24]

Elder Widtsoe's words—*something missing, incomplete, confusion, out of focus, static, semi-darkness and chaos*—describe the spiritual twilight. The Holy Ghost is the tool for dispersing the mists of darkness, for hearing and interpreting spiritual messages. Elder Widtsoe continues: "They who think the path difficult, have not tried it. . . . All who yield . . . obedience to God's law undergo a real transformation, by the Holy Ghost, which enables them more and more, to receive and understand spiritual messages. Unless that transformation is accomplished, a person is opaque to spiritual truth, and the 'things of God' are beyond his understanding. Great is the effect of such spiritual communication. . . . It transforms life. It makes the weak strong, the strong mightier. . . . The individual becomes filled with light as the incandescent lamp when the electric current passes through it. . . . [A]ll will be aided in their life pursuits if they have contact with the inexhaustible intelligence of the spiritual realm. The wealth of eternity will be theirs. They who do not seek to make themselves receivers of spiritual messages, but thrash about for such truth as their unaided powers may reveal, do not learn the meaning and destiny of life, and fail to win the vision of the glory of the universe in which we live."[25]

We might well ask, How many pure and holy things are waiting for me if I will learn the lesson that the young Prophet Joseph learned? The mature Prophet could later teach, "satan Cannot Seduce us by his Enticements unles we in our harts Consent & yeald—our organization such that we can Resest the Devil If we were Not organized so we would Not be free agents."[26] Since light and truth form the basis of our creation, we exist in a twilight only when we're living against our true nature—only when we're accepting lies. If a person really wants to discern between these two spirits, win the contest against evil, and escape that deadly spiritual twilight, he need only make the decision to do it, for, "Behold, my brethren, it is given unto you to judge, that ye may know good from evil; and the way to judge is as plain, that ye may know with a perfect knowledge, as the daylight is from the dark night" (Moroni 7:15).

Notes

This is a previously unpublished paper.

1. Oliver Cowdery, "Letter VIII," *Times and Seasons* 2, no. 13 (1 May 1841): 378.
2. *Words of Joseph Smith*, 62.
3. Ibid., 68.
4. *Gospel Doctrine*, 13.
5. Brigham Young, in *Journal of Discourses*, 13:145.
6. Brigham Young, *Discourses of Brigham Young*, comp. John A. Widtsoe (Salt Lake City: Deseret Book Co., 1978), 55.
7. *Words of Joseph Smith*, 200.
8. Ibid., 113–14.
9. Young, *Discourses of Brigham Young*, 55–56.
10. *Teachings of the Prophet Joseph Smith*, 157.
11. Brigham Young, in *Journal of Discourses*, 3:371.
12. Erastus Snow, in *Journal of Discourses*, 19:275–76.
13. Ibid.
14. Brigham Young, in *Journal of Discourses*, 11:14.
15. Brigham Young, in *Journal of Discourses*, 7:238.
16. Ibid., 238–39.
17. See *Teachings of the Prophet Joseph Smith*, 103.
18. Brigham Young, in *Journal of Discourses*, 7:26; emphasis added.
19. George Q. Cannon, in *Journal of Discourses*, 11:29–30.
20. Ibid., 30.
21. Heber C. Kimball, in *Journal of Discourses*, 10:240.
22. George Q. Cannon, in *Journal of Discourses*, 11:29–31.
23. Brigham Young, in *Journal of Discourses*, 3:371.
24. John A. Widtsoe, *Evidences and Reconciliations*, ed. G. Homer Durham (Salt Lake City: Bookcraft, 1960), 86–88.
25. Ibid., 88.
26. *Words of Joseph Smith*, 65.

"The Doer of Our Deeds"

The concept of the pursuit of self-esteem as a solution to man's most basic spiritual, emotional, and physical needs is not found in the scriptures. One might be surprised to find that in the scriptures there are no positive references to self-esteem, self-confidence, or self-love. This is not because God does not feel exquisite tenderness for human beings, but because he knows a better way for human flourishing than focusing on raising self-esteem.

Whatever the valid uses of the term *self-esteem* are, however much good is intended by it, I wonder if self-esteem isn't a red herring. The term *red herring* comes from the practice of dragging this smelly fish across a trail to destroy the original scent. Thus a red herring is a diversion intended to distract attention from the real issue. I suggest that the issue of self-esteem is a diversion to distract us from the real issue of our existence.

We might be justified in telling people to raise their self-esteem in order to solve their most basic problems if we knew nothing of man's premortal life or the spiritual purpose of his earthly probation and reductions or his glorious destiny. But the fulness of the gospel of Jesus Christ teaches the true nature and true needs of the human self and suggests that one cannot really define man without God.

I suggest that there are two major human conditions that the self is subject to that may have led to the idea that the pursuit of self-esteem was important: (1) man's vulnerability, or even pain, incident to the fall of man; and (2) the conflict or pain created by ignorance of divine law or by personal sin.

First, the pain incident to fallenness. Like our Savior, though to a lesser degree, we condescended to come to a fallen world, having agreed to submit to a considerable reduction in our premortal powers and quality of life. As we came to earth, separated from the presence of heavenly parents, we died spiritually (Helaman 14:16) and, in a sense, we were "orphaned." And now, with memory veiled and much reduced from our premortal estate, somewhat as aliens in a world that is inimical to our spiritual natures, we may carry an insecurity, a self-pain, which pervades much of our emotional life. Like Adam and Eve, we feel our self-consciousness or spiritual nakedness. The scriptures teach about this nakedness as a feeling of guilt or shame (see, for example, 2 Nephi 9:14 and Mormon 9:5). Do we have, as well, a sense of loss from deeply buried memories of who we once were in contrast with who we are now? But here are my main questions: Is it possible that in our efforts to find security we have fallen into a number of errors? Is it possible that we have created the whole issue of self-esteem in an attempt to soothe this fallen, homesick self?

There is a better way to find what our hearts long for than by seeking greater self-esteem. Our Savior, who felt all this pain himself (Alma 7:11–13), would not send us to earth without compensation for the distresses he knew we would feel, separated from him. He would not leave us comfortless. Recall the passages in John in which the Savior told the Twelve that he would be with them only a little while (John 13:33). Peter responded with "Lord, why cannot I follow thee now? I will lay down my life for thy sake" (v. 37). Jesus, sensing their pain, almost their desperation, at his leaving them, promised, "I will not leave you comfortless: I will come to you" (John 14:18). The English word *comfortless* translates the Greek word for *orphans:* "I will not leave you orphaned." The Savior continues, "If a man love me, he will keep my words: and my Father will love him, and we will come unto him, and make our abode with him" (v. 23). "My peace I give unto you: not as the world giveth, give I unto you. Let not your heart be troubled, neither let it be afraid" (v. 27).

Here we grasp the stunning insight that the Lord Jesus Christ himself is that consolation, that compensation, designed from the foundation of the world to comfort the human pain of fallenness, to compensate men and women for their earthly reductions and sacrifices. Only the Atonement, or more expressly the at-one-ment, of the Redeemer and the redeemed can heal the pain of the Fall. When we feel how much he loves us, we cannot help but love him: "We love him," John writes, "because he first loved us" (1 John 4:19). His love is the consolation.

Now to the second source of pain. The Lord explained, speaking to Adam: "When [thy children] begin to grow up, sin conceiveth in their hearts, and they taste the bitter" (Moses 6:55). What is this bitterness? The Lord says it is the conception of sin in our hearts. The pain of fallenness, then, is compounded by the bitterness of sin.

To understand why sin produces this bitterness, we remember that each individual spirit was begotten by glorious heavenly parents and thereby inherits a nature which, at its very core, is light, truth, intelligence, and glory (D&C 93:23, 29, 36). "Knowest thou not," the prophet John Taylor wrote, "that thou art a spark of Deity, struck from the fire of His eternal blaze, and brought forth in the midst of eternal burning?"[1] Christ says, "I am the true light that is in you, and . . . you are in me; otherwise ye could not abound" (D&C 88:50). Christ is the *life* and the *light* of every person (John 1:4, 9). King Benjamin teaches similarly that God preserves us from day to day, lending us breath, that we may live and move . . . even supporting us from one moment to another (Mosiah 2:21), and that all we have and are that is good comes from him (Mosiah 4:21).

I ask, if we live and move and have our being in him (Acts 17:28), where is self-esteem? How do I even separate my self out from the abundant grace that makes my life and even my intellect go forward in some marvelous symbiosis with my Creator? The human self cannot be defined without putting God in the definition.

It seems obvious that we—created out of the very stuff of truth, and permeated by his power—cannot live against our own natures of light and truth and intelligence without setting up conflict and spiritual dis-ease within ourselves. Sin goes against our most essential nature (Alma 41:11; Helaman 13:38). The quality of our emotional and spiritual existence is governed by divine law, and whether or not we know about these laws, or observe them, we are continually and profoundly affected by the laws of light and truth. Much of our unhappiness is self-inflicted through ignorance or through deliberate sin.

So here we have a challenging situation: a person, whose primeval nature is truth and light and purity, begins, under the influence of a fallen environment and a fallen body, to act against his spiritual nature. His sins of ignorance or choice produce bitterness, and he begins to suffer, but usually he doesn't know what the real source of his unhappiness is. He thinks it has something to do with the people around him, or he thinks it has to do with his circumstances. We don't entirely blame him for his confusion, because of course the reason we came to earth was to learn to discern good from evil so that we could be delivered from the miserable consequences of evil and darkness (2 Nephi 2:26). But, as Elder Neal A. Maxwell observes, "The heaviest load we feel is often from the weight of our unkept promises and our unresolved sins, which press down relentlessly upon us."[2]

Resistance to our spiritual natures manifests itself as guilt, despair, resentment, self-pity, fear, depression, feelings of victimization, fear over the scarcity of needed things, and other forms of distress. These are all functions of the fallen self and we all necessarily experience them. However, the pursuit of self-esteem will not solve the problems of the self that is in conflict because of sin or even of ignorant neglect of spiritual law.

There is another form of pain inflicted on people. Many people, especially children, are victimized by others. Their suffering is great and they often need much tender help and instruction in order to recover. Their understanding of who they are must be restored to them; they must accept what the Lord Jesus Christ is offering them.

Both of these healing truths are founded in faith and are processed through the Spirit. They must come to draw on the resources of truth and light, uncovering their own divinity and divine connections. But the pursuit of self-esteem will not solve the problems of those who suffer from others' sins against them.

When people are sinned against they often adopt ways of thinking and acting to defend themselves as they cope with their unhappy situations. They may learn to lie, or to try to manipulate and control others, or to live against their own deepest feelings of right and wrong, to blame, to resent, to resort to angry confrontations, and so on. These coping behaviors are understandable—they were developed in order to survive, but that doesn't lessen the truth that these behaviors are self-defeating and ultimately increase pain. Until a person stops sinning, no matter what the justifications for his sinning, no matter how innocently he began, he cannot get entirely well (Alma 41:15). In such a case, the sinned-against person must at some point acknowledge that he made choices that increased the pain of his victimization. The beauty of that realization is that a person can make new choices in greater harmony with light and truth. Elder Richard G. Scott spoke on being healed from the evil acts of others:

"No matter what the source of difficulty and no matter how you begin to obtain relief—through a qualified professional therapist, doctor, priesthood leader, friend, concerned parent, or loved one— no matter how you begin, those solutions will never provide a complete answer. The final healing comes through faith in Jesus Christ and His teachings, with a broken heart and a contrite spirit and obedience to His commandments.

" . . . Do what you *can* do a step at a time. Seek to understand the principles of healing from the scriptures and through prayer. . . . Above all, *exercise faith in Jesus Christ.*

"I testify that the surest, most effective, and shortest path to healing comes through application of the teachings of Jesus Christ in your life."[3]

The precepts of man cannot produce comprehensive healing and at best can endure only for a season (3 Nephi 27:11). Only the actual conversion of the fallen self through the power of the Lord Jesus Christ can rectify what is really amiss in a human being. An angel called this fallen self the natural man, saying, "The natural man is an enemy to God, and has been from the fall of Adam, and will be, forever and ever, unless he yields to the enticings of the Holy Spirit, and putteth off the natural man and becometh a saint through the atonement of Christ the Lord, and becometh as a child, submissive, meek, humble, patient, full of love, willing to submit to all things which the Lord seeth fit to inflict upon him, even as a child doth submit to his father" (Mosiah 3:19).

Could this putting off of the natural man through the Lord Jesus Christ actually be a recovery of our true, premortal self in all its wholeness and beauty, its love, its fearlessness and power?

We have the account of King Benjamin's people, who, upon hearing the word of God, became painfully conscious of their carnal state. They cried out, "O have mercy, and apply the atoning blood of Christ that we may receive forgiveness of our sins" (Mosiah 4:2). With that cry, their sensitive souls were cleansed by the Holy Spirit, top to bottom, of all their accumulations of willfulness, disobedience, and enmity; and into that vacuum rushed the sublime love of God. They received "peace of conscience, because of [their] exceeding faith . . . in Jesus Christ" (Mosiah 4:3—is it possible that "peace of conscience" is the Lord's term for what we call self-esteem?). Perhaps these Saints had not realized just how spiritually sluggish they were until that mighty power consumed in love all their sins and their pain and their sickness and their infirmity (Alma 7:11–13). They became acquainted with God's goodness and tasted his love. Not only that, they had come to their true self.

King Benjamin, seeing their joy, taught them how to retain it: "I would that ye should remember, and always retain in remembrance, the greatness of God, and your own nothingness, and his goodness and long-suffering towards you. . . . If ye do this ye shall always rejoice, and be filled with the love of God, and always retain a remission of your sins" (Mosiah 4:11–12).

What does the Lord mean by the *nothingness* of man? Several scriptures describe men in this way. King Benjamin asks, "Can ye say aught of yourselves? . . . Nay," he says (Mosiah 2:25). "Remember . . . your own nothingness" (Mosiah 4:11). Moses exclaims, "Now, . . . I know that man is nothing" (Moses 1:10). Ammon says, "I know that I am nothing" (Alma 26:12). Alma teaches that "man had fallen [and] could not merit anything of himself" (Alma 22:14). Nephi exhorts us to rely wholly on the merits of Jesus Christ (2 Nephi 31:19). Moroni speaks of the members of the Church "relying alone upon the merits of Christ, who was the author and the finisher of their faith" (Moroni 6:4). The Savior himself declares to his apostles, "Without me ye can do nothing" (John 15:5).

We recoil at *nothingness* because we try so hard to overcome our feelings of unimportance. But nothingness does not mean valuelessness. The Lord assures us that we are each of infinite worth to him. Rather, nothingness refers to man's fallen and reduced *state* in this mortal sphere (Mosiah 4:5). Nothingness describes not man's lack of value but rather his reduced powers during his mortal probation and, especially, his all-encompassing need for the Lord. Nothingness reminds us of the reductions we voluntarily subscribed to before the foundations of this world in order to come to earth and learn how to be taught from on high.

That is, nothingness is not only our actual state in the mortal probation, but also our personal perception when we persist in living in spiritual death, separate in our own minds from God. The Holy Ghost reveals that we are actually one with God (D&C 88:49–50) and when we align ourselves with the mind and will and power of God, that separation is healed and we *feel* alive in Christ (2 Nephi 25:25). The restoration of the rest of our self, through communion with God, fills us with the reality of fulness that we seek. Coming to Christ actually heals a false perception of separateness, which healing restores the fulness of all that each of us is.

Elder Richard G. Scott tells of a sacred experience in which strong impressions came to him during a period when he struggled to do a work the Lord had given him that was far beyond his personal capacity to fulfill. The Lord said to him, "'Testify to instruct, edify and lead others to full obedience, not to demonstrate anything of self. All who are puffed up shall be cut off.'" And then, "'You are nothing in and of yourself, Richard.' That was followed with some specific counsel on how to be a better servant."[4]

Ammon joyfully described his own nothingness: "I do not boast in my own strength, nor in my own wisdom; but behold, my joy is full, yea, my heart is brim with joy, and I will rejoice in my God. Yea, I know that I am nothing; as to my strength I am weak; therefore I will not boast of myself, but I will boast of my God, for in his strength I can do all things" (Alma 26:11–12). For Ammon, it seems, the whole concept of self-esteem was irrelevant. Being filled with the love of God was of far greater worth than any sense of self-confidence. If one grand objective of earth life is to gain access to the grace of Jesus Christ for our trials and divine development, we will immediately realize that self-confidence is a puny substitute for God-confidence. With respect to confidence, the Lord says, "Let thy bowels . . . be full of charity towards all men, . . . and let virtue garnish thy thoughts unceasingly; then shall thy confidence wax strong in the presence of God" (D&C 121:45).

Both Nephi and Mormon teach that when man is without charity he is nothing (2 Nephi 26:30; Moroni 7:44). Here we realize that as true followers of the Lord Jesus Christ we must embrace the gift of charity, which is the gift of happiness, in order that we might pass from nothingness to godliness. I am suggesting that we might want to substitute for the pursuit of self-esteem the pursuit of full discipleship with its attendant spiritual gifts, among which is the sublime spiritual gift of the pure love of Christ. Mormon wrote: "Wherefore, my beloved brethren, pray unto the Father with all the energy of heart, that ye may be filled with this love, which he hath bestowed upon all who are true followers of his Son, Jesus Christ; that ye may become the sons of God; that when he shall appear we shall be like him, for we shall see him as he is; that we may have this hope; that we may be purified even as he is pure" (Moroni 7:48).

The Lord identifies love and virtue as the essential ingredients in feelings of confidence and security. By these we dwell safely in the

Holy One of Israel (1 Nephi 22:28). Indeed, might not the pursuit of self-confidence actually pull us away from the connection the Lord is trying to make? Might it not merely produce carnal security (2 Nephi 28:21)?

One might notice that the pursuit of self-esteem seems to generate anxiety, while increasing humility and faith and connectedness with the Lord produce consolation and rest. Mormon describes Church members who, waxing "stronger and stronger in their humility, and firmer and firmer in the faith of Christ," are filled "with joy and consolation" (Helaman 3:35). Alma instructs his son to teach the people "to humble themselves and to be meek and lowly in heart; . . . for such shall find rest to their souls" (Alma 37:33–34).

Some may not like the dichotomy between the pursuit of self-esteem and faith in the Lord. Some may say that you can pursue and have both. But I do not find this idea of both pursuits in the scriptures; perhaps that is because there may be an incompatibility between the two. King Benjamin says, "Remember your own nothingness and God's goodness." In trying to have both, is there a possible double-mindedness? James says that "a double minded man is unstable in all his ways" (James 1:8). If the pursuit of self-esteem can lead to self-promotion, a person would be in spiritual trouble. Nephi says of self-promotion: "Priestcrafts are that men preach and set themselves up for a light unto the world, that they may get gain and praise of the world; but they seek not the welfare of Zion. Behold, the Lord hath forbidden this thing; wherefore, the Lord God hath given a commandment that all men should have charity, which charity is love. And except they should have charity they were nothing" (2 Nephi 26:29–30).

Here Nephi seems to view setting oneself up for a light to the world in order to get praise as being directly antithetical to having the pure love of Christ. One apparently can't do both. The Savior says, "Therefore, hold up your light that it may shine unto the world. Behold I am the light which ye shall hold up" (3 Nephi 18:24). Again he says that if our eye be single to his glory, our whole bodies will be filled with light: "Therefore, sanctify yourselves that your minds become single to God, and the days will come that you shall see him" (D&C 88:67–68). It seems as though the less attention we can give to self-esteem, the more light we can have.

Low self-esteem is often associated with feelings of incapacity, or a sense of victimization, or the realization that we can't make happen the opportunities, the approval, and the feelings that we need. But our relief comes when we realize that God has limited our powers so that as we cleave to him he can work his mighty miracles in our lives. Indeed, Moroni teaches that hopelessness and despair come from lack of faith in one's access to the Lord Jesus Christ (Moroni 10:22–23).

We may think that we or some other mortal opens the necessary doors to our future, but this conclusion is an error. We ourselves do not open these doors; only the Lord does. We exercise our agency through our choices, we make possible his miracles by our choices, but he retains the power to open or close the doors.

Often doors have closed before us that seemed to lead to the opportunities we thought we had to have. We may have assumed that the closed door was a reflection of some inadequacy in ourselves; but perhaps the closed door had nothing to do with whether we were good or bad or capable or incompetent. Rather, even now a loving Father shapes our path according to a prearranged, premortal covenant (Abraham 2:8); the opening or the closing of these various doors is dependent on the Lord's perfect perception of our developmental needs. All the elements that we really need for our individual experience here, he puts onto our path. The most important things that will happen to us in this life will come to us often by no initiative of our own, but rather because he is piloting the ancient plan. He says that he does nothing save it be for the benefit of the world (2 Nephi 26:24); he has promised that, if we will be faithful, all things will work together to our good in order that we may be conformed to the image of his Son (Romans 8:28–29).

With respect to doors closing or other kinds of divine deprivations, Elder F. Enzio Busche says, "When you are compelled to give up something or when things that are dear to you are withdrawn from you, know that this is your lesson to be learned right now. But know also that, as you are learning this lesson, God wants to give you something better."[5]

Therefore, we do not need to fear that our future lies in the fact that an authority over us plays favorites, or that a person's employer isn't well disposed toward him. Under such a belief, one might be tempted to think that only self-promotion, or image manipulation, or the compromising of what one really believes will open the doors. But even though someone in authority thinks he controls doors, there is really only one Keeper of the Gate (2 Nephi 9:41). "No weapon that is formed against thee shall prosper," he says. "This is the heritage of the servants of the Lord" (3 Nephi 22:17).

Now, I ask you, as various doors open and close, as the Lord Jesus Christ orchestrates even the details of our lives, as we are obedient to him, where is the *need* to pursue self-esteem? We don't need it. Faith in the Lord Jesus Christ will take us so much farther.

Christ himself is our model where the self is concerned. He says of himself:

1. "The Son can do nothing of himself, but what he seeth the Father do; for what things soever he doeth, these also doeth the Son likewise" (John 5:19).

2. "I do nothing of myself; but as my Father hath taught me, I speak these things" (John 8:28).

3. "The words that I speak unto you I speak not of myself: but the Father that dwelleth in me, he doeth the works" (John 14:10).

Moroni wrote that the resurrected, perfected Christ spoke to him in "plain humility" (Ether 12:39). Elder Neal A. Maxwell observed that "the Savior—the brightest individual ever to walk this planet—never sought to 'prosper' or to 'conquer' 'according to his genius' and 'strength'!" (Alma 30:17).[6] "Every man," Korihor taught, "fared in this life according to the management of the creature; therefore every man prospered according to his genius, and . . . every man conquered according to his strength" (Alma 30:17). Alma identified this precept that man prospers solely by his own resources as the doctrine of the anti-Christ.

248 · *Principles in Practice*

It seems to me that the fallen self may actually be an interloper in most of what we do and that we can find relief from the stresses and strains of self-promotion by saying, in effect, "Get thee behind me, Self." I wonder if this is what the Savior means when he says, "He who seeketh to save his life shall lose it; and he who loseth his life for my sake shall find it" (JST, Matthew 10:34). The fallen self seems to be a constant intruder as we strive for selflessness. President Ezra Taft Benson pointed out that "Christ removed self as the force in His perfect life. It was not *my* will, but *thine* be done."[7]

I have become aware of how demanding of attention the self is. What a lot of prayer and deliberate living and fresh insight it will take for me to remove my self as the force in my life! I have become aware that all my sins rise out of the self-absorption of my heart—impulses rising like the ticking of a clock in their persistent quest for self-gratification, self-defense, and self-promotion. It seems as though a change is needed at the very fountain of my heart, out of which all thought and emotion rise. Could I actually come to the point where I could act without calculating my own self-interest all the time? Could I really live my daily life so that I was constantly listening for the Lord's will and drawing down his grace to accomplish it? And when the Lord in his mercy meshes his power with my agency and my effort and brings forth some measure of success, I ask, where is self-esteem? Where is even the *need* for self-esteem? I feel as though I just want to say instead, "Lord, Increase my faith" (Luke 17:5).

How then does one appropriately think about oneself? I offer you Elder F. Enzio Busche's remarks. He said: "A disciple of Christ is . . . constantly, even in the midst of all regular activities, striving all day long through silent prayer and contemplation to be in the depths of self-awareness to keep him in the state of meekness and lowliness of heart."[8]

It seems appropriate as well to be conscious of our preciousness to our Father, while at the same time to feel meek and lowly before his sacrifices on our behalf, his respect for us, and his continuing graciousness to us. Again, Elder Busche spoke of the point at which we realize the Lord's love: "This is the place where we suddenly see the heavens open as we feel the full impact of the love of our Heavenly Father, which fills us with indescribable joy. With this fulfillment of love in our hearts, we will never be happy anymore just by being ourselves or living our own lives. We will not be satisfied until we have surrendered our lives into the arms of the loving Christ, and until He has become the doer of all our deeds and He has become the speaker of all our words."[9]

When Christ is the doer of all our deeds and the speaker of all our words, I have to ask, Where is *self*-esteem? I propose that self-esteem becomes a nonissue for the person who is perfecting his faith in the Lord Jesus Christ.

If I decide to give up some of the attention my self demands, what will I replace it with? The Lord answers, "Look unto me in every thought; doubt not, fear not" (D&C 6:36). The self is so demanding that perhaps one can only let go of the pursuit of self-promotion as one cleaves to the Lord Jesus Christ (Omni 1:26). As with Peter walking on the water, it may be our sudden self-consciousness that will cause us to fall (Matthew 14:28–30).

The world speaks of self-image, but Alma spoke of receiving the image of God in our countenances (Alma 5:14). In fact, we are informed: "All those who keep his commandments shall grow up from grace to grace, and become . . . joint heirs with Jesus Christ; possessing the same mind, being transformed into the same image . . . even the express image of him who fills all in all; being filled with the fulness of his glory; and become one in him, even as the Father, Son, and Holy Spirit are one."[10]

It seems that the perception of the self as an entity separate from God will, under the right conditions, get thinner and thinner.

President Ezra Taft Benson pressed us to be "changed for Christ," "captained by Christ," and "consumed in Christ."[11] We might ask, What is it that must be consumed? Maybe it is our old concept of self—the one we have learned from the precepts of men. Is it possible that the pursuit of self-esteem might delay this mighty change? Indeed, what if one ceased defining self-esteem or justifying one's pursuit of it, and just ignored it? What if, instead, one just began to obey whatever divine instruction one was not obeying, to sacrifice whatever needed sacrificing, and to consecrate whatever one was holding back? What if one just set out to "seek this Jesus" (Ether 12:41)?

So many issues that revolve around the subject of self fade like the dew in the sun as one cultivates faith in the Savior. Without him, nothing else matters. No amount of self-esteem or of anything else can adequately fill the void.

It is possible for the self to insulate itself from the love of the Lord Jesus Christ and not know it or feel it. The Savior's love is realized only when we open ourselves to the Spirit of the Lord through prayer and obedience. Otherwise, the thing we crave most, the *experience* of God's love, remains a hidden mystery. But no matter who we are or what we have done, we can repent, and may with full assurance seek to be clasped in the arms of Jesus.

The model for man's flourishing is in the scriptures. There we learn that, by ourselves, without Christ in our lives, we will feel the sorrows of the uncomforted, natural man. But with the Lord Jesus Christ, we will flourish. One who practices faith in the Lord Jesus Christ, and teaches others to do so also, will find relief from the stresses and anxieties of the pursuit of self-esteem.

Notes

Revised from an address originally given at a BYU Devotional, December 7, 1993; later published in *Spiritual Lightening* (Salt Lake City: Bookcraft, 1996), 17–30.

1. John Taylor, "Origin and Destiny of Woman," in *The Vision,* comp. N. B. Lundwall (Salt Lake City: Bookcraft, n.d.), 145–46.
2. Neal A. Maxwell, "Murmur Not," *Ensign,* November 1989, 85.
3. Richard G. Scott, "To Be Healed," *Ensign,* May 1994, 9; emphasis in original.
4. Richard G. Scott, "Acquiring Spiritual Knowledge," address delivered at BYU Education Week, 17 August 1993, 12.
5. F. Enzio Busche, "Unleasing the Dormant Spirit," BYU Devotional address, 14 May 1996.
6. Neal A. Maxwell, "Out of the Best Faculty," Annual University Conference, Brigham Young University, 23–26 August 1993.
7. Ezra Taft Benson, "Cleansing the Inner Vessel," *Ensign,* May 1986, 6; emphasis in original.
8. F. Enzio Busche, "Truth Is the Issue," *Ensign,* November 1993, 25.

9. Ibid., 26.

10. *Lectures on Faith*, 5:2.

11. Ezra Taft Benson, "Born of God," *Ensign*, July 1989, 4.

WHEN WE COME TO OURSELVES

A woman once asked a prophet about her origin, object, and destiny. His answer addresses both genders; he said, "Knowest thou not that thou art a spark of Deity, struck from the fire of His eternal blaze, and brought forth in the midst of eternal burnings?"[1] We were begotten in the likeness and image of our heavenly parents, free, rational, independent intelligences, "born and matured in the heavenly mansions, trained in the school of love in the family circle, . . . amid the most tender embraces of parental and fraternal affection."[2] This knowledge of our divine origin and nature can transcend every earth-made event that has taught us to the contrary—if we are humble enough to accept it. When a person does accept things as they really are (Jacob 4:13), he sees himself as he really is. If he doesn't know who he really is, nothing else will matter. But when he knows, he is ready for a new life. What I want to talk about here is the inner divine person waiting to be revealed.

I'm going to talk to you as though you don't yet know who you are—as though your eyes aren't yet open. What I know is that you have wasted a lot of time, as I have, feeling inadequate, shrinking back a little when new opportunities—even ones that you wanted— were offered to you; never feeling that you had measured up or could. Many experiences that might have become sweet memories have soured because you didn't feel that you played your role well enough. You've clipped your own wings time and time again, even when your spirit longed to fly.

We can be so cruel to ourselves; we destroy ourselves with our negative self-talk. We see only the raw material, much of it yet undeveloped, and we block from our awareness many already godlike qualities in us. We throw a wet blanket over all this divine fire. We steadfastly ignore the fact that many of our goals have been accomplished by our exercising godlike qualities: sacrifice, discipline, creativity, perseverance, love, faith. We decline to acknowledge the divinity that is developing in us through the challenges that life has presented to us for this very purpose. And we are ready to argue—in a most ungodly way—with anyone who tries to tell us something to the contrary. This refusal to accept our developing divinity is a sort of blasphemy and a denial of the purposes of the plan of salvation; it diminishes the effectiveness of our service and inhibits our full surrender to the Lord Jesus Christ. There is no real settled contentment, no real enlarging of the soul to meet great opportunities and possibilities, until we find our own divine fire.

If a person knows four basic truths, and he knows them in his spirit, he can live a different life—the one God offers him. These four empowering ideas are:

• know in spirit who you are
• know in spirit why you in particular are here
• know in spirit who God is and the nature of the spiritual universe that waits to support you, and
• know in spirit that each person is a creator and that the power is in him to bring miracles of change into his life and the lives of others.

KNOW IN SPIRIT WHO YOU ARE

The full truth behind what I'm going to say here can't be written. It is a personal revelation that is felt in the spirit. The truth of who you are is already present in your innermost being, but some of life's experiences conspire to block that knowledge. One of the adversary's most ardent endeavors is to keep that knowledge from our conscious awareness. The Lord, on the other hand, has told us repeatedly both in scripture and by his Spirit who we really are: "Ye were also in the beginning with the Father; that which is Spirit, even the Spirit of truth; . . . [i]ntelligence, or the light of truth, . . . spirit," eternal element, and innocence (D&C 93:23, 29, 33, 38). "Ye are gods; and all of you are children of the most High" (Psalm 82:6). "Ye are the sons of the living God" (Hosea 1:10). "Is it not written in your law, I said, Ye are gods?" (John 10:34). "We are the offspring of God" (Acts 17:29). "The Spirit itself beareth witness with our spirit, that we are the children of God" (Romans 8:16). "The inhabitants [of the worlds] are begotten sons and daughters unto God" (D&C 76:24).

You can know the truth of these scriptures in your spirit. Everyone has two main avenues of learning: the avenue through the natural, finite, carnal, or, what is often termed the ego mind; and the avenue through one's spirit. The spirit is a more comprehensive and more accurate way of knowing things, especially those beyond the physical senses. It is our personal connection with the eternal world and is continuous with the spirit of prophecy and revelation (Alma 5:45–46) and is therefore the place where the Holy Ghost speaks to us. It is also the place where we retrieve truths already present in our inner being. The spirit in us does not see limitations; rather, it sees possibilities. It leads its life in the consciousness that divine forces, divine beings, and the ingredients for miracles are always present.

The ego mind, on the other hand, has no frame of reference for miracles. It is characterized by its perception of limitations, by its susceptibility to fear, temptation, unbelief, pride, and discouragement. This is the place of stewing and rationalizing, analyzing, doubting, and criticizing. That's not to say that the ego mind should or could be done away with, but that it must become the servant, not the master, of the spirit.

Man exists then in one of two dimensions, spirit mind or ego mind, and many of us live back and forth between these two. Learning to distinguish between the two makes it possible to live more in spirit and to enjoy the privileges of that dimension. Because he is primarily a spiritual being with spiritual senses, possessing the light of Christ, every person will have had experiences in his spirit and will know things in his spirit. The Church member who has a testimony has already experienced learning through his spirit, and most of us have had many spiritual knowings blended into our daily experiences. The spirit is the place of feeling and "seeing" the truth and knowing that miracles happen all the time.

The veil operates to protect us from full memory of who we were in the premortal life. "The knowledge of our former state has fled from us . . . and the veil is drawn between us and our former habitation. This is for our trial. If we could see the things of eternity, *and comprehend ourselves as we are;* if we could penetrate the mists and clouds that shut out eternal realities from our gaze, the fleeting things of time would be no trial to us, and one of the great objects of our earthly probation or testing would be lost. But the past has gone from our memory, the future is shut out from our vision and we are living here in time, to learn little by little, line upon line, precept upon precept."[3]

Nevertheless, the Holy Spirit can cause us to remember things normally veiled. Two apostles wrote of the awakening of premortal memories: "During [a person's] progress in the flesh, the Holy Spirit may gradually awaken his faculties; and in a dream, or vision, or by the spirit of prophecy, reveal, or rather awaken the memory to a partial vision, or to a dim and half-defined recollection of the intelligence of the past. He sees in part, and he knows in part; but never while tabernacled in mortal flesh will he fully awake to the intelligence of his former estate. It surpasses his comprehension, is unspeakable, and even unlawful to be uttered."[4] "In coming here, we forgot all, that our agency might be free indeed, to choose good or evil, that we might merit the reward of our own choice and conduct. But by the power of the Spirit, in the redemption of Christ, through obedience, we often catch a spark from the awakened memories of the immortal soul, which lights up our whole being as with the glory of our former home."[5] Those awakened impressions can remind us to be that premortal person now.

If you have a patriarchal blessing, it declares your lineage in the great house of Israel. This information tells you about yourself in the premortal world. You did not enter this premortal family organization by birth; rather, you were chosen into this lineage based on your faith in the Lord Jesus Christ, along with your demonstrated desire and capacity to be exalted. Your election was a great privilege.

Peter called the house of Israel a chosen generation, a royal priesthood, a holy nation, a "special possession" of God (1 Peter 2:9; see note 9f). "Therefore, thus saith the Lord unto you, with whom the priesthood hath continued through the *lineage* of your fathers—for ye are *lawful heirs,* according to the flesh, and have been hid from the world with Christ in God" (D&C 86:8–9).⁶ Paul wrote to the Ephesian Saints, referring to our election in the life before, "Blessed be the God and Father of our Lord Jesus Christ, who hath blessed us with all spiritual blessings in heavenly places in Christ: according *as he hath chosen us in him before the foundation of the world,* that we should be holy and without blame before him in love: having [foreordained] us unto the adoption of children by Jesus Christ to himself" (Ephesians 1:3–5). One thing this scripture means is that those invited into the house of Israel in the premortal world were foreordained to become part of the eternal family of Jesus Christ and our heavenly parents in an exalted state. This preordination is more influential in our lives than we know. It means that not only are there dynamic forces operating on both sides of the veil assisting us to fulfill that foreordination, but that also there is an inner drive working, urging us forward.

Let us consider now our destiny in the celestial kingdom and one scripture only which declares that the Lord's covenant people, Israel, were foreordained to "come forth in the first resurrection; . . . and shall inherit thrones, kingdoms, principalities, and powers, dominions, all heights and depths . . . ; and they shall pass by the angels, and the gods, which are set there, to their exaltation and glory in all things, as hath been sealed upon their heads, which glory shall be a fulness and a continuation of the seeds forever and ever. Then shall they be gods because they have no end; therefore shall they be from everlasting to everlasting, because they continue; then shall they be above all, because all things are subject unto them. Then shall they be gods, because they have all power, and the angels are subject unto them" (D&C 132:19–20).

When we came to earth and fell into spiritual death, we separated not only from our heavenly parents and former associations but also, in a sense, from our premortal self. But, all this having been foreseen, the at-one-ment of the Lord Jesus Christ makes it possible for us to re-embrace that bright spirit in this life. Simply speaking, the Lord wants us to know who we really are and helps us to reidentify with that vibrant premortal person. But the forces of spiritual death work against that reunion, as each of us could testify: "The 'real me,' or 'the spiritual child of God,' created in innocence and beauty, is engaged in a fight for life or death with the elements of the earth, the 'flesh,' which, in its present unredeemed state, is enticed and influenced by the enemy of God. . . . Without Christ, this war within us is lost. . . . We knew that before we came to this earth, and we can sense it again, when through the Light of Christ our minds are quickened with understanding."[7]

Here in our brief time on earth we have great opportunities for development. Elder Parley P. Pratt described the potential and the process: "An intelligent being, in the image of God, possesses every organ, attribute, sense, sympathy, affection that is possessed by God Himself. But these are possessed by man, in his rudimental state. . . . Or, in other words, these attributes are in embryo; and are to be gradually developed. They resemble a bud, a germ, which gradually develops into bloom, and then, by progress, produces the mature fruit, after its own kind."[8] That is, the mortal probation is for exercising the power of divinity, strengthening and developing the seeds of divinity that have lain in our spirit for eons. Our spirit is a conduit through which the Holy Ghost can gain access to these attributes and develop them: "The gift of the Holy Ghost adapts itself to all these organs or attributes. It quickens all the intellectual faculties, increases, enlarges, expands and purifies all the natural passions and affections; and adapts them, by the gift of wisdom, to their lawful use. It inspires, develops, cultivates and matures all the fine-toned sympathies, joys, tastes, kindred feelings and affections of our nature. It inspires virtue, kindness, goodness, tenderness, gentleness and charity. It develops beauty of person, form and features. It tends to health, vigor, animation and social feeling. It invigorates all the faculties of the physical and intellectual man. It strengthens, and gives tone to the nerves. In short, it is, as it were, marrow to the bone, joy to the heart, light to the eyes, music to the ears, and life to the whole being."[9] These are some of the beautiful things God seeks to do for man. The Holy Ghost can work within us and compensate for all or any of our perceived weaknesses. There are no restrictions, there are no limits to our ability to develop the very same attributes that he and the Savior have if we will open to a new vision.[10]

Commenting similarly on man's divine potential, President Spencer W. Kimball said, "Man can transform himself and he must. Man has in himself the seeds of godhood, which can germinate and grow and develop. As the acorn becomes the oak, the mortal man becomes a god. It is *within his power* to lift himself by his very bootstraps from the plane on which he finds himself to the plane on which he should be. It may be a long, hard lift with many obstacles, but it is a real possibility."[11] On another occasion he spoke of our innate holiness: "In each of us is the potentiality to become a god—pure, holy, influential, true and independent of all these earth forces. We learn from the scriptures that each of us has an eternal existence, that we were in the beginning with God. And understanding this gives to us a unique sense of man's dignity."[12]

Continuing, he reaffirmed the reality of man's potential for perfection: "I would emphasize that the teachings of Christ that we should become perfect were not mere rhetoric. He meant literally that it is the right of mankind to become like the Father and like the Son, having overcome human weaknesses and developed attributes of divinity. [Even though] many individuals do not fully use the capacity that is in them, [that] does nothing to negate the truth that they have the power to become Christlike. It is the man and woman *who use the power* who prove its existence; neglect cannot prove its absence."[13] "Man alone, of all creatures of earth, can change his thought pattern and become the architect of his destiny."[14]

It will be helpful to summarize briefly President Kimball's points by his key words: *transform, seeds of godhood; potentiality, pure, holy, influential, true, independent; dignity, right to become, capacity, use the power; change his thought pattern, architect of destiny.* These words describe the plan for every person.

Part of changing the thought pattern has to do with deeply accepting who one is, regardless of how earth experiences may have shaped a self-concept to the contrary. A limited self-concept is formed from misinterpretations of temporary earth events, but it ceases to have significance as it is viewed against the backdrop of our true, eternal being. Every single person has developable divinity—it doesn't matter where he has been or what he has done or how long he has ignored the powers in his soul. He cannot kill the divinity in his own soul. Rather, he can change his mind.

Is it our concept of humility that creates conflict with allowing our divinity to expand and express itself boldly and energetically? Maybe we think of humility as being caved in, self-effacing, self-demeaning, weak, ineffective, overly tolerant of our own and other's self-lies, setting artificial self-limits—and calling all these virtue. Alma instructed his son to be bold (though not overbearing) (Alma 38:12). Ammon, a potent personality, spoke boldly to the Lamanite king (Alma 18:24). But the boldness sprang from his knowing who he was and why he was there in front of the king, and it had nothing to do with his ego mind or ego objectives. He knew he was Christ's because he had made himself so by his choices. He had made the choice to act from the divinity within. Moroni described Jesus as humble even in his resurrected and glorified state (Ether 12:39), and yet, who could deny that the Lord's humility cohabits with his omnipotence? That is, humility can coexist with power and influence.

Humility must be something like this: knowing who you are and Whose you are and Whose power and will are moving through you. The divine nature is revealed, felt, and unfolds as a person accepts truth in his spirit and makes new choices based on that knowledge.

Sometimes just *experiencing* the truth changes a person, and he is no longer able to make the same old self-limiting choices.

KNOW WHY YOU ARE HERE

Knowing who you are points you to the next truth. Because of your origin and destiny, there are always higher purposes operating in your life and the lives of those around you. Many of these higher purposes are obvious, but some are more subtle, and some are completely unknown to you at this moment; but they are always operating, seeking to make you conscious of them so that you can respond appropriately. All of these higher purposes have to do with your life mission. Instead of resisting life's experiences as though they appeared randomly on your path, you can wake up and align yourself with what is happening in the moments of your life. Your life's experiences are organized for the accomplishing of your life mission.

Your mission is your path to personal fulfillment. It was arranged in the life before based on the things you wanted to learn and to contribute; it is written in your soul. So that when you are involved in what you personally came to earth to do, you may feel flashes of, "Ah, this is the reason I was born." Your mission will have things in common with many others, but the details will be personal, the particulars uniquely yours. You will know that you have come into this time and place both to learn and to make the contributions that are in your soul.

In discovering your own mission, the scriptures, your patriarchal blessing, and the desires of your heart all provide significant clues. In addition, the calls that come to you from the Church are designed to help you to fulfill a work that you covenanted to do before this life. Here are some mission indications from the scriptures that will pertain to some degree to your own. Note what they have in common:

"For the fulness of mine intent is that I may persuade men to come unto [Christ]" (1 Nephi 6:4).

Alma is sealed up to eternal life because of his desire to gather the Lord's sheep (Mosiah 26:19–20).

"O Lord, . . . spare my life, that I may be an instrument in thy hands to save and preserve this people" (Alma 2:30).

"They fasted much and prayed much . . . that they might be an instrument in the hands of God to bring . . . the Lamanites, to the knowledge of the truth" (Alma 17:9).

"I will make an instrument of thee in my hands unto the salvation of many souls" (Alma 17:11).

"This is the blessing which hath been bestowed upon us, that we have been made instruments in the hands of God to bring about this great work" (Alma 26:3; see also v. 15).

"This is my glory, that perhaps I may be an instrument in the hands of God to bring some soul to repentance; and this is my joy" (Alma 29:9).

Our admittance into premortal Israel was a response to our desire to consecrate ourselves to awakening others to the real purposes of our life.

We are never alone in our personal mission; it always has to do with other people who are dependent on us—very often without knowing it—to fulfill our role. Premortal covenants were made. People, living, dead, and yet unborn, depend on our living in alignment with our higher purposes. We are each an essential link in the great design for the family of God. God made each of us important in the interconnectedness, as well as the progress, of mankind. In fact, as a member of the house of Israel, you were foreordained to participate in lifting mankind temporally and spiritually.

Elder J. Reuben Clark, referring to the Church's welfare plan, made a statement that could serve as part of a personal mission statement: "The real long term objective of the Welfare Plan is *the building of character* in the members of the Church, givers and receivers, *rescuing all that is finest deep down inside of them,* and bringing to flower and fruitage *the latent richness of the spirit,* which after all is the mission and purpose and reason for being of this Church."[15]

Because not only you, but also each person you know, is a

storehouse of abilities and resources waiting to be developed, you are called to be a rescuer of all "that is finest," "the latent richness of the spirit," in those around you. This mission empowers them, and therefore empowers you. As to the particulars of your own life mission, if you really want to know—you will know.

WHO GOD IS

"Now, this is the truth. We humble people, we who feel ourselves sometimes so worthless, so good-for-nothing, we are not so worthless as we think. There is not one of us but what God's love has been expended upon. There is not one of us that He has not cared for and caressed. There is not one of us that He has not desired to save and that He has not devised means to save. There is not one of us that He has not given His angels charge concerning. We may be insignificant and contemptible in our own eyes and in the eyes of others, but the truth remains that we are children of God and that He has actually given His angels . . . charge concerning us, and they watch over us and have us in their keeping."[16] These words not only tell us about us, but they describe the true nature of our Heavenly Father. "If men do not comprehend the character of God they do not comprehend themselves,"[17] because God is the loving and merciful developer of the divinity in his children. The plan of salvation requires the intimate involvement of our loving Heavenly Father. "As a man through the powers of his body he could attain to the dignity and completeness of manhood, but could go no further; . . . through the *essence and power of the Godhead, which is in him,* which descended to him as the gift of God from his heavenly Father, he is capable of rising from the contracted limits of manhood to the dignity of a God, and thus through the atonement of Jesus Christ and the adoption he is capable of eternal exaltation, eternal lives and eternal progression. *But this transition from his manhood to the Godhead can alone be made through a power which is superior to man*—an infinite power, an eternal power, even the power of the Godhead: for as in Adam all die, so in Christ *only* can all be made alive."[18]

That is, we can never fully unfold without the personal attention of the Lord Jesus Christ and our Heavenly Father (see, e.g., Moroni 9:26). We can sense their intimate and monitoring presence. They know what we most need and whisper it to us, often in the moment: "And thine ears shall hear a word behind thee, saying, This is the way, walk ye in it" (Isaiah 30:21). Indeed, man's design allows him to contain the presence of God. Consider these scriptures:

"Man is the tabernacle of God, even temples" (D&C 93:35).

"If a man love me, he will keep my words: and my Father will love him, and we will come unto him, and make our abode with him" (John 14:23).

"The light shineth in darkness, and the darkness comprehendeth it not; nevertheless, the day shall come when you shall comprehend even God, being quickened in him and by him. Then shall ye know that ye have seen me, that I am, and *that I am the true light that is in you,* and that *you are in me;* otherwise ye could not abound" (D&C 88:49–50; emphasis added).

Christ comes to us not only to awaken and enlarge our divinity but also simply to let us experience him, to have communion with him for our mutual blessing. "Behold, I stand at the door, and knock: if any man hear my voice, and open the door, I will come in to him, and will sup with him, and he with me" (Revelation 3:20). The power of Christ is in you; it is seeking expression and expansion. But you have to own it. You have to give yourself permission to have Christ and to be the person God has created you to be.

Three examples illustrate how people can accept the Lord's invitation and grow beyond their dreams under his nurturing hand: Moses called on the name of the Lord, and beholding his glory, heard a voice, saying: "Blessed art thou, Moses, for I, the Almighty, have chosen thee, and thou shalt be made stronger than many waters; for they shall obey thy *command as if thou wert God.* And lo, *I am with thee,* even unto the end of thy days" (Moses 1:25–26). "Enoch . . . bowed himself to the earth, before the Lord, . . . saying: Why is it that

I have found favor in thy sight, and am but a lad, and all the people hate me; for I am slow of speech; wherefore am I thy servant?" (Moses 6:31). The Lord answered, "Behold my Spirit is upon you, . . . *and thou shalt abide in me, and I in you;* therefore *walk with me*" (Moses 6:34). On the occasion of Jeremiah's learning that he had been foreordained to be a prophet in the premortal world (Jeremiah 1:5), he said, "Ah, Lord God! behold, I cannot speak: for I am a child. But the Lord said unto me, Say not, I am a child. . . . Be not afraid of their faces: for *I am with thee* to deliver thee, saith the Lord. Then the Lord put forth his hand, and touched my mouth. . . . Behold, I have put my words in thy mouth" (Jeremiah 1:6–9). "For all flesh is in my hands" (Moses 6:32).

MEN AND WOMEN ARE CREATORS

Among the attributes of God are his desire and his ability to create. As children of God, by our divine genetics, we are creators through choice, faith, and committed action. We initiate a process by our desires; focused, persistent desires and committed action call forth a response from a dynamic universe. When our powers of creation are connected to our faith in the Lord Jesus Christ, they become very effective. Elder Packer described the creative function of faith: "There is another kind of faith, rare indeed. This is the kind of faith that *causes* things to happen. It is the kind of faith that is worthy and prepared and unyielding, and it calls forth things that otherwise would not be. It is the kind of faith that moves people. It is the kind of faith that sometimes moves things. . . . It comes by gradual growth. It is a marvelous, even a transcendent, power, a power as real and as invisible as electricity. Directed and channeled, it has great effect."[19]

Sometimes Latter-day Saints think that godhood will happen one day as if by magic, "'that in the end, when the day comes that the Lord will make them gods or goddesses, when someone lays their hands on their heads and, as it were, says to them, You have now all that you need to be a God—go ahead—this is not true. All that you need to be a God is in you right now. Your job is to take those crude elements within you and refine them.'"[20] As Elder Gene R. Cook commented, "In other words, the Lord is saying, 'Take the reins. Take charge under the direction of my Spirit. Don't wait for someone to tell you everything to do.' . . . 'For *the power is in them . . .*' (D&C 58:26–28; italics added.)"[21]

In order to take the reins, one must understand the comprehensive power of choice. Without choice there is no existence, because, by the nature of intelligence, each has agency (D&C 93:30; 2 Nephi 2:11–12). In its most primitive state, an intelligence begins to make choices, each of which is self-evolutionary. Choice is powerfully creative because there are no choices without consequences (Alma 41:15). Each choice increases or decreases the scope of action of an intelligent being. Our own intelligence was created with full potential to become godlike, but the actual development moves forward by the nature of the choices made.

Every human being has the power to choose and indeed chooses his reality every minute by what he activates within his own soul. If we have dealt with negative ways of looking at things or with a sense of being victimized by people or circumstances, this has been a choice, and it creates its own reality. Then in our spiritual ignorance, our distorted emotions and toxic feelings make us feel that thinking, believing, or acting a different way *wouldn't make any difference* in our lives. But the truth is that we move from dimension to dimension based on what we choose to believe and do. States of victimization are self-chosen; states of self-liberation are also self-chosen. We are capable of activating different realities—like degrees of glory—in this life, and depending on the laws we neglect or observe, we ascend toward a state of blessedness which can be tasted in this life, or we descend to a degree of misery, which is very common.

No one can activate a different reality for another person; that person must do it for himself. The Lord has been blessing us all along and has given us everything we need to activate a better reality, but we haven't believed him. We may have searched books looking for The Insight that will change everything in a moment. We wait for something miraculous and utterly transforming to happen to us instead of accepting the miraculous nature of our developing being, believing the Lord, and just implementing divine principles. Miracles are happening all the time.

The spirit of this process is captured in some counsel given by President Harold B. Lee: "If you want the blessing, don't just kneel down and pray about it. Prepare yourselves in every conceivable way you can in order to make yourselves worthy to receive the blessing you seek."[22] Paul put the principle this way: "I have planted, Apollos watered; but God gave the increase [or *growth* of the seed]" (1 Corinthians 3:6). That is, bringing to pass through faith that which does not now exist is a creative combination of our absolute reliance on divine truth, our persisting desire, and our committed action—which the Lord causes to prosper—if it be according to his will. But I hasten to add, there is very much out there to create which is already in harmony with the Lord's will, which is already in alignment with the higher purposes operating in your life, and which is yet undone. We live below our privileges where creating is concerned.

Our spirit has been encouraging us to face the truth of our divine being for a long time. It may even have precipitated breakdowns of various kinds in order to press us to that insight. Even so, we may have buried the demands—or the call to truth—of our spirit in our subconscious, where these truths war with our actual choices. It is a war to see which spirit, of truth or of lies, will have ascendance in our souls.

"This war is a war that has to be fought by all of Heavenly Father's children, whether they know about it or not. But without a keen knowledge of the plan of salvation, and without the influence of the divine Light of Christ to bring us awareness, this war is being fought subconsciously, and therefore its battlefronts are not even known to us, and we have no chance to win. Wars in the inner self that are fought subconsciously, with unknown battlefronts, lead to defeats which also hurt us subconsciously. These defeats are reflected in our conscious life as expressions of misery, such as a lack of self-confidence, lack of happiness and joy, lack of faith and testimony, or as overreactions of our subconscious self, which we see then as pride, arrogance, or in other forms of misbehavior—even acts of cruelty and indecency.

"No! There is no salvation without Christ, and Christ cannot be with us unless we pay the price of the constant fight for self-honesty."[23]

To deny our divinity by our thoughts, words, and behavior is to have to keep lying to ourselves and to others. To be who we really are is to give a gift to others.

Until we come to ourselves, we are likely to be self-focused in a negative sense. We have a lot of energy tied up in trying to establish our own worth and we're missing the coming and going of many higher purposes trying to get our attention. We free up a lot of energy getting settled on who we are, why we're here, who God is, and how our own choices affect our reality.

The whole issue reduces to our willingness to change our minds. Like a computer, your subconscious mind is already programmed, but can be reprogrammed if you don't like the way it's serving you. You and I have absolute stewardship over our mind. We can change the contents of our mind. We can put into it what we want and what we put in has creative power over our soul. The more we think a thought, the deeper it is embedded in the mind. The more we accept the revelations of the Lord and gently but persistently focus on who we really are and why we are really here, the more we evolve. You can simply see yourself as one with your eternal history, a repository of godly powers. If you change your perception of yourself by a willingness to accept new truth and you feed your mind with life-changing ideas, you change yourself and your circumstances.

Jesus said, "If thou canst believe, all things are possible to him that believeth" (Mark 9:23). "In a world filled with skepticism and doubt, the expression 'seeing is believing' promotes the attitude, 'You show me, and I will believe.' . . . When will we learn that in spiritual things it works the other way about—that believing is seeing?"[24] Eternal principles held in the mind become not only the focus of our lives, they become the medium through which the power of Christ works to produce the mighty change. But we must supply the medium through which the Spirit is to work.

IMPLICATIONS

So now we look at the implications of accepting who we are and why we're here. One might start in a small degree to bring to pass the desires of his heart. What do I want to create in my life? "Even as you desire of me so it shall be unto you; and if you desire, you shall be the means of doing much good in this generation" (D&C 6:8). "If you will ask of me you shall receive; if you will knock it shall be opened unto you" (D&C 6:5). A sense of awakened goodness seeks an avenue to accomplish good things, "for the power is in them, wherein they are agents unto themselves" (D&C 58:28). Ultimately all our best desires will have to do with seeking "to bring forth and establish the cause of Zion" (D&C 6:6) in all its temporal and spiritual implications—because we are premortal Israel come to earth for the great cause of Jesus Christ.

What does a revelation of one's divinity mean for daily life? Among many possibilities, I'll list the following: It means that we will take risks every day to be who we really are in order to fulfill our particular purposes and mission here. It means that we will clarify our intention, commit ourselves to specific action, and take our place among those who—worthy, prepared and unyielding—move people and things to be better than they would otherwise have been. It means being true to the inner divinity by telling the truth and keeping one's word so that we don't have inner conflict over promises made in vain. It means that we listen for the divine voice that prompts us at least daily from the fountain of our being.

It means that since we acknowledge God as the primary force now in our lives, no one can enlarge or diminish us by words or actions, because all the beauty, passion, power, and love in us are eternal. It means, being a giver of life to others, that we empower other people, inspiring their excellence; we know in our sensitivity to higher purposes that their inner spirit also longs to expand and break the chains of self-limitation (see, e.g., Alma 5:7, 9).

Here at the close of this chapter, will you join with me in your imagination for a minute? The human imagination is, after all, the rudimentary organ of creation. What we imagine can come to be. I'm facing a vision of three spheres: premortal world on the left, planet Earth in the middle, and the celestial kingdom on the right. You have seen something of who you were and what you achieved in the premortal world, as well as the kingdom for which you are headed. But what is happening to you, that same beautiful spirit, in this middle world? Can you unite all three beings into the one *you* to bless you now? The Prophet Joseph said that "the past, the present, and the future were and are, with [the Lord], one eternal 'now.'"[25] Therefore, we are already one with what will be. And can you, at the same time, sense the presence of the Lord whose history and love for you have been intertwined with yours from the beginning? It is a matter of willingness. Therefore, let believing become seeing.

Notes

This is a previously unpublished paper.

1. John Taylor, "Origin and Destiny of Woman," in *The Vision*, comp. N. B. Lundwall (Salt Lake City: Bookcraft, n.d.), 145–46.
2. Parley P. Pratt, *Key to the Science of Theology*, 10th ed. (Salt Lake City: Deseret Book Co., 1965), 57.
3. Charles Penrose, in *Journal of Discourses*, 26:28; emphasis added.
4. Pratt, *Key to the Science of Theology*, 58.
5. *Gospel Doctrine*, 13–14.
6. The emphasis in this scripture and in subsequent scriptures in this chapter has been added by the author.
7. F. Enzio Busche, "Truth Is the Issue," *Ensign*, November 1993, 24.
8. Pratt, *Key to the Science of Theology*, 100–101.
9. Ibid., 101.
10. Adapted from Richard L. Millett, "Letter to Missionaries," 1 March 1997; unpublished document in possession of author.
11. Spencer W. Kimball, *The Teachings of Spencer W. Kimball*, ed. Edward L. Kimball (Salt Lake City: Bookcraft, 1982), 28; emphasis added.
12. Ibid., 26.
13. Ibid.; emphasis added.

14. Ibid., 27.
15. J. Reuben Clark, Jr., in *A Leader's Guide to Welfare: Providing in the Lord's Way* (Salt Lake City: The Church of Jesus Christ of Latter-day Saints, 1990), inside front cover; emphasis added.
16. George Q. Cannon, *Gospel Truth,* 2 vols., comp. Jerreld L. Newquist (Salt Lake City: Deseret Book Co., 1974), 1:2; emphasis added.
17. *Words of Joseph Smith*, 340.
18. John Taylor, *The Mediation and Atonement* (Salt Lake City: Deseret News Publishing Co., 1892), 141; some emphasis added.
19. Boyd K. Packer, "What Is Faith?" *Faith* (Salt Lake City: Deseret Book Co., 1983), 42–43; emphasis in original.
20. James E. Talmage, as quoted by Gene R. Cook, *Living by the Power of Faith* (Salt Lake City: Deseret Book Co., 1985), 91; emphasis in original.
21. Cook, *Living by the Power of Faith*, 91–92.
22. Harold B. Lee, *Stand Ye in Holy Places* (Salt Lake City: Deseret Book Co., 1974), 244; see also 2 Nephi 25:23.
23. Busche, "Truth Is the Issue," 25.
24. Packer, "What Is Faith?" 43.
25. Joseph Smith, *History of The Church of Jesus Christ of Latter-day Saints,* 7 vols., 2nd ed. rev., ed. B. H. Roberts (Salt Lake City: The Church of Jesus Christ of Latter-day Saints, 1932–51), 4:597.

.

Scripture Index

OLD TESTAMENT

NEW TESTAMENT

BOOK OF MORMON

1 NEPHI

1:1, p. 15
1:12, p. 52
1:14, p. 13
1:20, p. 62
2:12, p. 210
2:20, pp. 66, 211
3:7, p. 220
3:29, p. 73
4:3, p. 62
5:4, p. 105
5:17, p. 52
6:4, p. 261
7:12, p. 13
7:21, p. 200
8:10, p. 77
9:6, pp. 13, 176
10:12–14, p. 77
11:7, 8, 21–22, 25, p. 77
12:19, p. 224
13:19, p. 63
13:23, 25–27, 29, p. 112
13:24–29, p. 120
13:27, p. 118
15:16, p. 78
15:22, p. 77
15:36, p. 78
16:28–29, pp. 56, 67
17:3, 30, p. 216
17:23–31, p. 65
17:36, p. 21
17:46, p. 13
18:11, p. 212
20:10, p. 216
21:14–16, p. 216
22:28, pp. 14, 245

2 NEPHI

1:15, pp. 80, 99
2:11–12, p. 266
2:15, p. 171
2:26, p. 240
2:27, p. 219
2:29, p. 230
4:17–19, p. 70

4:21, p. 89
4:27–33, pp. 64, 200
4:28, p. 231
4:28–31, 35, p. 71
5, p. 200
6:17, p. 63
7:2, p. 63
9:1–12, 19, p. 62
9:5–7, p. 8
9:10–13, p. 63
9:14, p. 238
9:20, p. 13
9:41, p. 247
9:52, p. 220
10:23, p. 220
25:23, pp. 174, 271
25:25, pp. 111, 243
25:27, p. 112
26:24, pp. 14, 176, 246
26:29–30, p. 245
26:30, p. 244
27:3, p. 223
27:10, p. 13
28:20–22, p. 229
28:21, p. 245
31:9, p. 151
31:19, pp. 7, 243
32:3, pp. 53, 73
32:5, p. 53

JACOB

2:5, p. 13
4:6, p. 52
4:11, p. 78
4:12, p. 75
4:13, p. 253
4:14, pp. 75, 80, 118
4:17–18, p. 75
4:18, p. 185
5, p. 75
5:18, p. 79
5:47, 75, p. 81
5:61, 74, p. 78
6:3, 4–7, p. 81
6:8, p. 79

ENOS

1:15, p. 63

OMNI

1:7, p. 63
1:8, p. 73
1:26, p. 249

WORDS OF MORMON

1:13–14, 18, p. 88

MOSIAH

1:11, pp. 88, 94, 96
1:11–12, p. 91
2:4, 34, p. 89
2:9, p. 84
2:21, p. 239
2:25, p. 243
2:32, p. 196
3:3–4, p. 90
3:4, pp. 90, 91
3:6, pp. 219, 232
3:13, p. 90
3:16, p. 17
3:18, 25, 26, p. 91
3:19, pp. 82, 143, 171, 242
4:1–2, p. 102
4:2, p. 242
4:3, pp. 89, 91, 242
4:5, p. 243
4:11, pp. 90, 243
4:11–12, p. 242
4:12, 20, p. 91
4:13, p. 194
4:21, p. 239
5:2, pp. 92, 94
5:4, p. 90
5:5, p. 94
5:7, pp. 90, 93
5:15, p. 95
7:27–28, p. 161
7:29, pp. 80, 170, 211
7:33, pp. 64, 170
8:18, p. 86

ENOS (cont.)

12:31–32, p. 112
18:10, p. 143
18:21, p. 196
21:15, p. 69
23:21–22, p. 212
23–24, p. 69
24:12–16, p. 70
24:13–14, p. 15
26, p. 29
26:19–20, pp. 30, 262
27:22, p. 187
27:29, p. 227
27:30, pp. 221, 227
29:19–20, p. 65

ALMA

2:30, p. 262
5:6–9, p. 224
5:7, 9, p. 270
5:14, p. 249
5:26, 28, p. 92
5:37–38, pp. 80, 169
5:38–39, p. 228
5:39, p. 217
5:45–46, p. 255
7:11–13, pp. 238, 242
8:15, p. 173
9:10, p. 62
10:6, p. 172
12:29, p. 90
13, pp. 19, 156
13:1–3, 6, p. 23
13:3, pp. 8, 9, 19
13:4, pp. 22, 27
13:5, 16, p. 22
13:10–12, p. 29
13:12, pp. 93, 149, 193
13:17–18, p. 88
13:18, p. 149
13:28, p. 229
17:2–3, p. 52
17:9, 11, p. 262
18:24, p. 260
22:14, p. 243

DOCTRINE AND COVENANTS

PEARL OF GREAT PRICE

Subject Index

Abomination of desolation, 41–42
Abraham, 23–24, 99, 149
Abuse, 200, 240–42
Adam, 99, 227, 229
Adonai, 117
Adversity, 151–55, 210, 211–13, 238–40. *See also* Trials
Afflictions, 15, 16, 201
Agency: of man, 14, 266; in mortality, 171–73, 175; of children, 183–84
Alcoholism, 179–81
Allegories, understanding, 75
Alma the Elder, 29–30, 66, 69–70
Alma the Younger: on premortal valiance, 9, 23; on sanctification, 29; conversion of, 29–30, 71–72, 227; understands instruments of revelation, 54–55, 67–68; on deliverance of God, 62–64; on being spiritually reborn, 92, 249
Ammon, 244, 260
Amulek, 172
Angels, 90, 147–48, 202–3
Anger, 181
Anointing, 78–79
Anthropomorphisms, 116–17
Anthropopathisms, 116–17
Anti-Christ, 247
Anxiety, 185–87
Apostasy, Jewish: after Malachi, 109; nature of, 110–12; continuation of, 120–21; clouds doctrine, 141

Apostles' Creed, 126
Aquinas, 128
Arethusa, Mark, 127
Armor of God, 178
Artaxerxes I, 111
Asceticism, 115, 157–61
Atonement (or at-one-ment): deliverance symbolic of, 72–73; olive tree symbolic of, 76–82; personal nature of, 90–91; effectiveness of, altered, 161; consumes imperfections, 177; works of, 189–92; living spirit of, 197, 204–6; power of, thwarts Satan, 219; heals pain of Fall, 239; helps reidentify premortal selves, 258. *See also* Relationships
Augustine, 128, 132, 158

Babel, Tower of, 100
Ballard, Melvin J., 9
Baptism: for the dead, 125, 138–41; in early Christian churches, 131–32
Bath-qol, 118–20
Bede, 128, 132
Benjamin, King: priesthood of, 84–86; establishes righteous community, 88; people of, are spiritually reborn, 89–94, 102, 242; people of, receive name of Christ, 94–96; teachings of, 239, 245
Benson, Ezra Taft: on death, 16; on

Bonus Chapter 1

The Provocation in the Wilderness and the Rejection of Grace

M. Catherine Thomas

Camped in the hot, waterless wilderness of southern Palestine, the Israelites challenged Moses, saying, "Wherefore is this that thou hast brought us up out of Egypt, to kill us and our children and our cattle with thirst?" (Exodus 17:3). This complaint might have been understandable had these people never seen the hand of God in their lives, but this incident occurred after the miraculous Passover, after their passage through the Red Sea dry shod, and after the outpouring of manna and quail from heaven. In response to the Israelites' faithlessness, an exasperated Moses cried out to the Lord, "What shall I do unto this people? they be almost ready to stone me" (Exodus 17:4). The Lord answered: "Behold, I will stand before thee there upon the rock in Horeb; and thou shalt smite the rock, and there shall come water out of it, that the people may drink. And Moses did so in the sight of the elders of Israel. And he called the name of the place Massah, and Meribah" (Exodus 17:6–7).

Psalm 95 provides the linguistic link that identifies this incident as the Provocation: "To day if ye will hear his voice, Harden not your heart, as in the *provocation* [Hebrew *meribah*], and as in the day of temptation [Hebrew *massah*] in the wilderness: When your fathers tempted me, proved me, and saw my work. Forty years long was I grieved with this generation, and said, It is a people that do err in their heart, and they have not known my ways: Unto whom I sware in my wrath that they should not enter into my rest" (Psalm 95:7–11; emphasis added; see also Hebrews 3:8–11, 15).

The event at Meribah is the Provocation mentioned throughout the Bible. In that incident, the Lord tested the faith of the children of Israel and their willingness to accept His love and grace. Grace is the Lord's divine enabling power, given to humankind to help them with all the challenges of their lives; grace ultimately empowers them to lay hold on heaven itself. But the Israelites' response to the Lord's abundant generosity illustrates a religious paradox: God offers His children grace, but the children will not seek it; God offers His children heaven, but the children will not enter in.

We shall see that the Provocation refers not only to the specific incident at Meribah but to a persistent behavior of the children of Israel that greatly reduced their spiritual knowledge (see Psalm 95:10: "they have not *known* my ways"; emphasis added) and thus removed them from sublime privileges. After a succession of provocations, the Israelites in time rejected and lost the knowledge of the anthropomorphic nature of the Gods, the divine relationship of the Father and the Son, and the great plan of grace inherent in the doctrine of the Father and the Son.

The Israelites sought to be self-prospering and became angry when the God of Israel tested or tried them. The Provocation constitutes a recurring theme in the Old Testament, and indeed, in every extant scripture since. The pages of Exodus and Deuteronomy, which narrate the history of the Israelites in the wilderness, describe three additional incidents of provocation. First, at the foot of Sinai, where the Lord tried to sanctify His people and to cause them to come up the mountain, enter His presence, and behold His face, the Israelites refused to exercise sufficient faith to overcome their fear and enter into the fire, smoke, and earthquake that lay between them and the face of God. They said to Moses, "Speak thou with us, and we will hear: *but let not God speak with us,* lest we die" (Exodus 20:19; emphasis added). Moses responded, "Fear not" (Exodus 20:20). Nevertheless, "the people stood afar off, and Moses drew [alone] near unto the thick darkness where God was" (Exodus 20:21).

Second, when the Israelites were camped at Kadesh Barnea in the wilderness, the Lord tried to bring them into the promised land, but they were so frightened by the report of giants in the land that neither Moses nor Caleb and Joshua could get them to exercise enough faith to enter and conquer the land (see Deuteronomy 9:22–23). Again, as at Massah and Meribah, they refused the grace of the Lord.

Third, again at Sinai, when Moses went up to receive the fulness of the gospel from the Lord on the first set of plates, the Israelites made and set up the golden calf. Their rejection of the Lord in the very moment that Moses was receiving the fulness of the gospel for them was a most serious provocation. When he discovered what they had done, Moses broke the tables before the children of Israel. A second, lesser set of plates was made, but they were missing "the words of the everlasting covenant of the holy priesthood" (Joseph Smith Translation, Deuteronomy 10:2), meaning the higher, sanctifying ordinances of the Melchizedek Priesthood. Those were the very ordinances that gave access to the presence of the Lord (see Joseph Smith Translation, Exodus 34:1–2).

With their rejection of the higher priesthood, Israel began to lose the true doctrine of the Father and the Son.[1] The Lord gives the reason: "This greater priesthood administereth the gospel and *holdeth the key* of the mysteries of the kingdom, *even the key of the knowledge of God.* Therefore, in the ordinances thereof, the power of godliness is manifest. And *without the ordinances thereof,* and the authority of the priesthood, the power of godliness is not manifest unto men in the flesh; For without this no man can see the face of God, even the Father, and live. Now this Moses plainly taught to the children of Israel in the wilderness, and sought diligently to sanctify his people that they might behold the face of God; But they hardened their hearts and could not endure his presence; therefore, the Lord . . . swore that they should not enter into his rest while in the wilderness, which rest is the fulness of his glory. Therefore, he took Moses out of their midst, and the Holy Priesthood also" (D&C 84:19–25; emphasis added).

The Prophet Joseph Smith observed: "God cursed the children of Israel because they would not receive the *last law* from Moses. . . . When God offers a blessing or knowledge to a man and he refuses to receive it he will be damned. . . . The Israelites [prayed] that God would speak to Moses [and] not to them in consequence of which he cursed them with a carnal law. . . . [The] law revealed to Moses in Horeb . . . never was revealed to the [children] of Israel."[2] Thus, the children of Israel wandered an unnecessary forty years in the wilderness as God tried to teach them to rely on Him.

We really begin to appreciate the Old Testament when we realize that Israel's experiences in the wilderness are both literal and allegorical of our own experiences. Moses, speaking of manna as a symbolic teaching device, said, "[God] humbled thee, and suffered thee to hunger, and fed thee with manna . . . *that he might make thee know* that man doth not live by bread only, but by every word that proceedeth out of the mouth of the Lord" (Deuteronomy 8:3; emphasis added).

The Apostle Paul spoke similarly of the manna and the water and the rock: "Brethren, I would not that ye should be ignorant, how that all our fathers were under the cloud, and all passed through the sea; And were all baptized unto Moses in the cloud and in the sea; And did all eat the same *spiritual meat* [manna]; And did all drink the same *spiritual drink* [water at Meribah]: for they drank of that *spiritual Rock* that followed them: and that Rock was

Christ" (1 Corinthians 10:1–4; emphasis added). The Savior called Himself manna, or the Bread of Life (see John 6:51, 54), indicating mankind's persisting need for divine nourishment.

Exploring scriptural symbols further, in both the Old Testament and the Book of Mormon a *wilderness* symbolizes any place in which the people are tested, tried, proven, refined by trials, taught grace, and prepared to meet the Lord (see Alma 17:9; see also Christ's preparations in the wilderness in Matthew 4:1–2). Scriptural *journeys* often symbolize man's earthly walk from birth through the spiritual wildernesses of a fallen world (see Ether 6:4–7 for the ocean allegory of man's journey; see also 1 Nephi 8 for the *path* leading to the tree of life). God seeks to teach that His children cannot be self-prospering and thereby fulfill the purposes of their earthly lives. They must learn to seek and accept His grace to reach their destinations, which are *promised lands* or places of deliverance and spiritual peace where Zion can be established. The Lord speaks to modern Israel: "Zion cannot be built up unless it is by the principles of the law of the celestial kingdom; otherwise I cannot receive her unto myself. And my people must needs be chastened until they learn obedience, if it must needs be, by the things which they suffer" (D&C 105:5–6). Therefore, the Lord provides in our lives wildernesses and waterlessness and overwhelming challenges to entice His children to involve Him as they struggle through life.

The Book of Mormon supplies further insight into what the Provocation actually refers to. Jacob referred to Psalm 95 (on the plates of brass) when he wrote: "Wherefore we labored diligently among our people, that we might persuade them *to come unto Christ,* and *partake* of the goodness of God, that they might *enter into his rest,* lest by any means he should swear in his wrath they should not enter in, as in the *provocation* in the days of temptation while the children of Israel were in the wilderness" (Jacob 1:7; emphasis added).

Alma enlarged the implications still further in speaking of the first provocation, or man's first spiritual death at Adam's fall, and the second provocation, or man's continuing spiritual death that comes through rejecting the Lord: "If ye will harden your hearts ye shall not enter into the rest of the Lord . . . as in the *first provocation,* yea, according to his word in the *last provocation.* . . . Let us repent, and harden not our hearts, that we provoke not the Lord our God . . . but *let us enter into the rest of God,* which is prepared according to his word" (Alma 12:36–37; emphasis added).

The Provocation, then, seems to encompass a preference for spiritual death—a preference for a return to Egypt—rather than the demanding trek through repentance to sanctification. The Provocation, in all its manifestations, implies a refusal to come to Christ to exercise faith in the face of such a daunting call, a refusal to partake of the goodness of God, a refusal to accept the restoration to God's presence or rest, a refusal to allow the Savior to work His mighty power in one's life, a refusal to enter into the at-one-ment for which He suffered and died, a refusal to be "clasped in the arms of Jesus" (Mormon 5:11). The Provocation is anti-Atonement and anti-Christ. Abinadi laments over men and women who have "gone according to their own carnal wills and desires; having never called upon the Lord while the arms of mercy were extended towards them; for the arms of mercy were extended towards them, and they would not" (Mosiah 16:12).

But who, indeed, was the God who had stood before Moses upon the rock at Meribah? (see Exodus 17:6–7). That God had revealed Himself to our fathers Abraham, Isaac, and Jacob as a glorified, exalted man, that is, as an anthropomorphic (in the form of man) God who had created male and female in the image of heavenly parents. [3] This God sought a constant interaction with and a response from His children. He spoke of Himself as father and Israel as His children (see Malachi 2:10). He spoke of the covenant people as bride and Himself as bridegroom (see Hosea 2:19–

20). The scriptures ring with manlike descriptions of an interactive God: "The eyes of the Lord" (Psalm 34:15), the ears of the Lord, and the mouth of the Lord; the heavens as the works of His fingers (see Psalm 8:3); the tablets of the covenant "written by the finger of God" (Exodus 31:18). We read of "his countenance" (Numbers 6:26), which He causes to shine or which He hides. We read of His "right hand" (Psalm 118:16), His arm stretched out in mercy and invitation. In Genesis He walks about in the garden (see Genesis 3:8), He goes down to Sinai or to His temple (see Genesis 11:5; 18:21) to reveal Himself (see Exodus 19:18, 34:5) and to dwell in the midst of the children of Israel, and He goes up again (see Genesis 17:22; 35:13). He sits on a throne (see Isaiah 6:1) and causes His voice to be heard among the cherubim (see Numbers 7:89). Moses not only sees the Lord's back (see Exodus 33:23) but speaks to Him face to face and mouth to mouth (see Numbers 12:8). Among several emotions, the Lord expresses tenderness, mercy, love, joy, delight, and pity, as well as sadness, frustration, and anger.

With the loss of the Melchizedek Priesthood, however, and the Jews' resulting vulnerability to Greek and other cultural and philosophical influences, there arose among the Jews a resistance to the idea of an anthropomorphic God. At least by the intertestamental period (the period following Malachi, between the Old Testament and the New), the scribes and rabbis found the anthropomorphisms in the Hebrew Bible offensive and made small textual changes, which they described as "biblical modifications of expression" [4] (see Jacob 4:14 for Jacob's acknowledgment of Israel's deliberate mystification of God). For example, in place of "I [God] will dwell in your midst," they substituted "I shall cause you to dwell," avoiding the idea that God would dwell with men. The text of Exodus 34:24 was subtly altered from "to see the face of the Lord" *(lir'ot 'et-pene yhwh)* to the phrase "to appear before the Lord" *(lera'ot 'et-pene yhwh)*. Again, the effect is to distance and dematerialize God.

It appears that the Jewish translators of the Septuagint Bible (from Hebrew to Greek; abbreviated LXX) also attempted to dematerialize God.[5] An example is Exodus 29:45 (KJV): "I will dwell among the children of Israel, and will be their God." Instead of "I will dwell," the Septuagint reads, "And I shall be called upon [or named] among the children of Israel and will be their God" (Exodus 29:45).[6] The effect of the change from *to dwell* to the phrase *to be called upon* is to distance God from His children.[7]

This attempt to dematerialize God is also found among the Israelite apostates in the Book of Mormon. Ammon and Aaron had to teach that God, the Great Spirit, would not always be spirit, but would tabernacle Himself in the flesh (see Alma 18:34–35; 22:8–14; see also Mosiah 3:5). Abinadi, in fact, was martyred for his very declaration that this spirit God would take on the form of man in order to perform the great Atonement (see Mosiah 13:32–35). The apostate Zoramites' belief that God is a spirit and never would be anything else really meant they believed there would be no Christ, no incarnation of God on earth, and thus, no Atonement (see Alma 31:15–16).

A Jewish scholar named Philo lived in the period just prior to Jesus' advent. His writings, which influenced Judaism as well as Christianity, taught that the physical and emotional references in the scriptures to God were allegorical, not literal. He wrote that when Moses described God with human emotions, the reader needed to know that "neither the . . . passions of the soul, nor the parts and members of the body in general, have any relation to God."[8] Philo explained that Moses used these expressions as an elementary way to teach those who could not otherwise understand. Thus, when the Savior came to the Jews in the meridian of time, He found many of them obsessed with religion, with purity, and with scrupulous observance of law, but He found few who knew God.

Removing the body, parts, and passions from God also removes His ability to suffer and thus obscures the real meanings behind the Atonement. The Book of Mormon, however, teaches that one of the reasons the Savior came to earth was His desire to "take upon him [mankind's] infirmities, that his bowels may be filled with mercy, according to the flesh, that he may know according to the flesh how to succor his people according to their infirmities" (Alma 7:12). Alma quotes Zenos on the accessibility of the Father's grace through the Atonement of the Son: "And thou didst hear me because of mine afflictions and my sincerity; and it is because of thy Son that thou hast been thus merciful unto me, therefore I will cry unto thee in all mine afflictions, for in thee is my joy; for thou has turned thy judgments away from me, because of thy Son" (Alma 33:11). Alma then quotes Zenock on the nature of Israel's Provocation: "Thou art angry, O Lord, with this people, because they will not understand thy mercies which thou hast bestowed upon them because of thy Son" (Alma 33:16).

Related to God's nature is God's name. The reluctance to offend God by anthropomorphic references grew stronger with time so that even the use of the name *YHWH* (Yahweh or Jehovah) was avoided. At least by the third century B.C., *adonai,* meaning "lord," was substituted for the divine name[9] and it ultimately became both illegal and blasphemous to speak the name aloud among the Jews, even in the temple or synagogue. One scholar notes: "The divine name, once the 'distinguishing mark' of divine presence and immanence, had become the essence of God's unapproachable holiness so that in the Jewish tradition 'the Name' (*ha shem*) could be synonymous with 'God.'"[10] A moment's reflection leads us to see that since God had ordained His name as a keyword by which a covenant person could gain access to Him (see Moses 5:8; 1 Kings 8:28–29; Mormon 9:21), to forbid the divine name was to forbid access, through holy ordinances, to God Himself.

With the dematerializing of God came the obscuring of the Father-Son relationship. Religious history reveals that one major apostate objective has been to merge the members of the Godhead into one nebulous being. That merging clouds several significant truths, among which I mention two in passing and a third for discussion:

1. The doctrine of a divine Father and Son begins to reveal that there must be family relationships, parents, husbands, and wives, all of which continue in the eternities.

2. Eternal families being possible, there is need for temple ordinances that seal these relationships for eternity.

3. The Son models for humankind the relationship of grace by which one gains exaltation and which men and women must model in order to be like the Gods.

It is particularly that last truth that I would like to explore here, but first a word about merging the Gods into one amorphous being. That which set the Israelites apart from all others in the polytheistic Greco-Roman and Near Eastern cultures was their steadfast declaration of one omnipotent God, that is, their belief in monotheism. It was perhaps because they had interpreted Deuteronomy 6:4, "Hear, O Israel: The Lord our God is one Lord" to mean that there was only one God, that the later Jews rejected Christ (see John 8:41, 58–59). After all, Christ taught that He is the Son of God, and so, they said, He made Himself equal with God and seemed, in fact, to be multiplying Gods (see John 5:18).[11] Nevertheless, although it is true that there is one omnipotent God, that truth is not the whole truth. When the Savior came to the earth in the meridian of time, one of His tasks was to restore the Melchizedek Priesthood and thus restore the *knowledge of the Father.* Jesus taught that He, the Son, is the only avenue to exaltation or reunion with the Father.

But one of the most important revelations from the divine Father- Son relationship is the model it provides of the *nature of a saving relationship* with God. The Savior showed us how to live in total submission; He drew continually on His Father's grace. He says, "The Son can do *nothing* of himself, but what he seeth the Father do: for *what things soever* he doeth, these also doeth the Son. . . . For the Father loveth the Son, and sheweth him all things that himself doeth" (John 5:19–20; emphasis added). And again, "As the living Father hath sent me, and *I live by the Father:* so he that eateth me [reference to the sacrament], even *he shall live by me*" (John 6:57; emphasis added). And again, "I do *nothing* of myself; but as my Father hath taught me, I speak these things. And he that sent me is with me: the Father hath not left me alone; for I do always those things that please him" (John 8:28–29; emphasis added). Further, He said, "Believest thou not that I am in the Father, and the Father in me? the words that I speak unto you I speak not of myself: but the Father that dwelleth in me, *he doeth the works*" (John 14:10; emphasis added). Ultimately, Christ will even deliver up the kingdom, for which He died, to His Father (see D&C 76:107). This relationship of the at-one-ment of the Father and the Son is the divine model for the Saints of God and was revealed that we might emulate it.

The Savior taught this at-one-ment relationship to His disciples and, indeed, to all who become His disciples. The means of at-one-ment with the Son and the Father is the Holy Ghost. It is through cultivating the Holy Ghost that we enter into at-one-ment with the Son and the Father. Jesus told His disciples: "I will pray the Father, and he shall give you another Comforter, that *he may abide with you* for ever; Even the Spirit of truth . . . for *he dwelleth with you,* and *shall be in you*" (John 14:16–17; emphasis added; see also John 17:20–23).[12]

The Apostle Paul experienced this relationship of oneness with the Savior; he wrote to the Galatians: "I am crucified with Christ: nevertheless I live; yet not I, but Christ liveth in me: and the life which I now live in the flesh I live by the faith of the Son of God, who loved me, and gave himself for me" (Galatians 2:20).

In scenes recorded in 3 Nephi, the resurrected, perfected Christ gave abundant evidence of His continuing dependence on His Father. He makes frequent reference to the commandments and will of His Father. He seems very eager to return to the full presence of His Father (3 Nephi 17:4); we see Him kneel and bow Himself to the earth, pouring out both His troubled heart (3 Nephi 17:14) as well as His joy (3 Nephi 17:20–21), His thanks (3 Nephi 19:20, 28), and His needs (3 Nephi 19:21, 29). Perhaps this relationship of divine dependence and atonement continues far into the eternities. It is revealed to us in this life so we can learn to live in that relationship and thus gain admission to that community of grace-linked Gods.

The relationship of grace helps us understand more fully this passage in Doctrine and Covenants: "[Christ] received not of the fullness at the first, but received grace *for* grace; And he received not of the fulness at first, but continued from grace *to* grace, until he received a fulness. . . . I give unto you these sayings that you may understand and know how to worship, and know what you worship, that *you* may come unto the Father in my name, and in due time receive of his fulness. For if you keep my commandments you shall receive of his fulness, and be glorified in me as I am in the Father; therefore, I say unto you, you shall receive grace *for* grace" (D&C 93:12–13, 19–20; emphasis added).

By this scripture we understand that as Christ gave grace to those around Him, He received from His Father increasingly more grace to give. Thus, receiving grace *for* grace, Jesus grew from grace *to* grace: a model for us. "Freely ye have received, freely give," the Savior told His disciples (Matthew 10:8). The Lord has blessed each of us individually many times over with many more forms of grace than we now know or could count. Perhaps all of the Lord's grace to us—His many kindnesses to each of us, our talents, our gifts of spirit and personality, our bodies, our material resources—is given to us so that we will have something to give one another. As we give of this grace in countless ways to those around us, especially where it may not seem to be merited, the Lord increases His gifts of grace to us; in this process of our receiving grace *for* the grace we give, we grow from grace *to* grace, as Christ did, until we obtain a fulness.

Living in such a relationship as the Father and the Son's, either on earth or in heaven, requires a total willingness to dethrone oneself as the regent in one's own kingdom and to enthrone Christ as He enthroned the Father. President Ezra Taft Benson observed that "Christ removed self as the force in His perfect life. It was not *my* will, but *thine* be done."[13]

How privileged we are to know about the relationship of grace and to know of the divine possibilities for ourselves through connection with the Father and the Son, to experience the exquisitely loving and personal nature of the Gods in their great chains of light and grace.

In various forms, the Provocation continues with us today. We recognize in ourselves the rejection of grace as we keep trying to struggle through life on our own judgment and power, keeping our own personal agenda on the throne. Mormon described the philosophy of the anti-Christ Korihor as the belief that man prospers and conquers by his own strength and genius, not through dependence on a greater divine being (see Alma 30:17). Thus, struggling alone without calling on God reflects the doctrine of the anti-Christ. It is apparent that even the Son of God could not have prospered without His Father's grace.

Moroni also emphasizes grace: "And now, I would commend you to *seek this Jesus* of whom the prophets and apostles have written, that the *grace of God the Father,* and also *the Lord Jesus Christ,* and the Holy Ghost . . . abide in you forever" (Ether 12:41; emphasis added; see also Moroni 10:32).

We see in Israel's provocations a key to understanding nearly every interaction between God and Israel recorded in the pages of the Bible. On the one hand, God's whole efforts are bent toward helping the covenant people to prosper through His grace; on the other hand, Israel strives to be self-prospering. In the midst of abundant miracles and divine gifts, the persistent rejection of God's grace is Israel's Provocation.

---

[1] Of course, all the prophets had the Melchizedek Priesthood, but their right to confer it or teach its mysteries was restricted. See Joseph Smith, *Teachings of the Prophet Joseph Smith,* comp. Joseph Fielding Smith (Salt Lake City: Deseret Book, 1938), 181.

[2] Andrew F. Ehat and Lyndon W. Cook, eds., *The Words of Joseph Smith*(Provo, UT: Religious Studies Center, Brigham Young University, 1980), 244, 247; emphasis added.

[3] Because God created man in His own image, it is more accurate to speak of man as *theomorphic* (in the form or image of God) than to speak of God as anthropomorphic.

[4] Cecil Roth and Geoffrey Wigoder, eds. *Encyclopedia Judaica,* (Jerusalem: Keter, 1982), 3:54, s.v. "anthropomorphism."

[5] Perhaps it is helpful to note here, with respect to apostate movements, that in any apostasy there are the deliberate initiators and perpetrators of lies (see 1 Nephi 13:27; Jacob 4:14; Moses 1:41), but there is usually also a larger group of innocent and well-intentioned victims (see 1 Nephi 13:29; D&C 123:12). Not all promoters of false ideas have malignant intent; most are to some extent the victims of those who have gone before.

[6] Another example is found in Numbers 12:8. The Hebrew version reads: "With [Moses] will I speak mouth to mouth, even apparently, and not in dark speeches; and the image [or *form*] of the Lord shall he behold." The Greek version reads: "Mouth to mouth will I speak to him, in his sight and not in riddles, and he shall see the glory of the Lord." The change from *image* to *glory* is from the specific to a more nebulous description of God.

[7] For a fuller discussion of the Apostasy of the doctrine of God during the intertestamental period, see the author's chapter entitled "From Malachi to John the Baptist: The Dynamics of Apostasy," *Studies in Scripture,* Vol. 4: *1 Kings to Malachi,* ed. Kent P. Jackson (Salt Lake City: Deseret Book, 1993), 471–83. See also an in-depth study of the dematerializing of God in the author's "The Influence of Asceticism on the Rise of Christian Text, Doctrine, and Practice in the First Two Centuries," (PhD diss., Brigham Young University, n.d.).

[8] Philo, "The Unchangeableness of God," *Loeb Classical Library*(Cambridge, MA: Harvard University Press, 1930), 3:37.

[9] Roth and Wigoder, *Encyclopedia Judaica,* 7:680, s.v. "God, Name of."

[10] Another scholar suggests that, with ascendancy of the law in Israel and the need to buffer the law against violations, *any* use of the divine name had to be denied. The prohibition was motivated by a desire to ensure that the name would not be used "in vain" (Exodus 20:7) either by Jews or non-Jews. The name used in the temple or the synagogue was eventually affected by this fear. In the Septuagint the name of Yahweh was rendered throughout with *kyrios* ("Lord"), following the Jewish preference for *adonai.* Martin Rose, "Names of God in the OT," *The Anchor Bible Dictionary* (New York: Doubleday, 1992), 1010.

[11] An extension of this merging of Gods occurred in the early period of the Christian Church at the Council of Nicea (AD 325) when the decision to fabricate a trinity of three beings into one made it possible to make Christianity securely monotheistic, again in a threateningly pagan environment. Both of these beliefs, monotheism and trinitarianism, did violence to the full truth about the true nature of the Godhead and of godliness itself (see my article, "The Conspiracy Begins," in *Selected Writings of M. Catherine Thomas: New Expanded edition,* M. Catherine Thomas [Salt Lake City: Digital Legend Press], 287-302).

[12] Joseph Smith, *Lectures on Faith*, comp. N. B. Lundwall (Salt Lake City: Bookcraft, n.d.), Lecture 5, 48–49, explains <u>how</u> the Father and the Son are one through the medium of the Spirit and how all the Saints may in the same manner come into at-one-ment with them: "The Only Begotten of the Father, full of grace and truth, and having overcome, received a *fullness* of the glory of the Father, *possessing the same mind* with the Father, which *mind is the Holy Spirit,* that bears record of the Father and the Son, and these three are one; or, in other words, these three constitute the great, matchless, governing, and supreme power over all things . . . the Father and the Son possessing the same mind, the same wisdom, glory, power, and fullness—filling all in all; the Son being filled with the fullness of the mind, glory, and power; or, in other words, the *spirit,* glory, and power, of the Father . . . which Spirit is shed forth upon all who believe on his name and keep his commandments. . . . all those who keep his commandments shall grow up from grace to grace, and become heirs of the heavenly kingdom, and joint heirs with Jesus Christ; possessing the same mind, being transformed *into the same image* or likeness, even *the express image* of him who fills all in all; being filled with the fullness of his glory, and become one in him, even as the Father, Son and Holy Spirit are one" (emphasis added).

[13] Ezra Taft Benson, in Conference Report, April 1986, 6.

———

The Conspiracy Begins

M. Catherine Thomas

> *Why should it be thought a thing incredible with you,*
> *that God should raise the dead?*
> ACTS 26:8

Two grief-filled nights passed into the early hours of the third day as the Lord's body lay entombed. But as the spring sky began to glow and birds began their morning songs, there was a stirring within the dark, sealed room. We know that it was the Redeeming Lord returning from His spirit ministry; heedless of the impenetrable walls of the sepulchre, He reentered and reclaimed the lifeless body. No mortal eye beheld the power of the Father raise the Son to become the first fruits of the resurrection (Romans 8:11; 2 Nephi 2:8). But soon after, the stone now rolled away and the tomb vacated, mortal witnesses to the greatest of all miracles would begin to abound. Soon the weeping Mary would turn and answer His greeting and behold Him walking in the garden, vibrant, radiant, alive. As first witness, she would not only see the reality of the resurrection but also hold Him.[1] Shortly afterward, the Lord would greet the other women

---

[1] JST John 20:17, "*Hold* me not"; that is to say, "*Detain* me not"; or, "*Stop clinging to me,*" cited in Arndt and Gingrich, *GEL*, s.v. "ᶜἅπτω," 126.

who would hold Him by the feet, worshipping Him. Then He would appear to the two disciples on the way to Emmaus, then to Peter, then to the Ten, then to the Eleven, and so the witnesses would multiply.

We know from the New Testament record that Peter saw the resurrected Lord on at least six separate occasions; his associates saw Him nearly that many. Luke reported that "he shewed himself alive after his passion by many infallible proofs, being seen of them forty days, and speaking of the things pertaining to the kingdom of God" (Acts 1:3). These appearances included not only sight and hearing but also the touching of the Lord's resurrected body and the witnessing of His interaction with the material world. One of the Lord's primary purposes for these appearances was to impress on those who had known Him most intimately that the dead live again in tangible, physical, durable bodies. Thus, they were made eyewitnesses who could testify of "that which was from the beginning, which we have heard, which we have seen with our eyes, which we have looked upon, and our hands have handled, of the Word of life" (1 John 1:1).

But the vast majority of the world neither saw nor touched the Lord's glorified body. And many since have found it difficult to believe in something they could not establish empirically. After evaluating all extant evidence about Jesus between about A.D. 30–200, one highly publicized Christian group found that Jesus was simply "a secular sage"; they concluded, "The Christ of creed and dogma, who had been firmly in place in the Middle Ages, can no longer command the assent of those who have seen the heavens through Galileo's telescope."[2] The group's founder declared that Jesus never asked us to believe that He would be raised from the dead, nor did He regard scripture as inspired.[3] Truly, it has been thought a thing incredible that God should raise the dead (see Acts 26:8).

---

[2] Robert Funk, R.W. Hoover, *The Five Gospels: The Search for the Authentic Words of Jesus* (New York: Macmillan, 1993), 2.

[3] Robert W. Funk, "The Gospel of Jesus and the Jesus of the Gospels," keynote address to the Jesus Seminar Fellows, in *The Fourth R* (Santa Rosa, California: Westar Institute, November/December 1993), 8.

My purpose here will be to analyze two main deterrents to finding the resurrection credible. The first deterrent is the belief that spiritual truth is not knowable. People have asked, "How, in the absence of present, tangible, empirical evidence, is the pursuit of the truth of the resurrection to go forward, relying only on the intangibles of faith? How is the seeker to distinguish the truth from the vain imaginations of men? How is it that those who have not seen and yet believe are more blessed than those who have seen?" (John 20:26–29; 3 Nephi 12:2). My premise is that unseen, spiritual truth is indeed knowable, and that even when key elements in it cannot be verified empirically, scripture is useful as a starting place for the pursuit of truth.[4]

The second deterrent is that Satan inspired people to devise various formal and informal plots to distort the truth about the resurrection of Jesus Christ. To uncover these plots, our study will lead us back into pre-Christian and early Christian history and reveal the pivotal nature of the doctrine of the Lord's resurrection. That is, encompassed in the doctrine of the resurrection are the issues that most threaten the enemies of truth.

What we will uncover here are the workings of a deliberate cover-up of the literal, bodily resurrection of Christ. We will examine the relationship of the resurrection to two main aspects of the conspiracy: (1) the plot against the body of man and the body of God and (2) the plot against man's oneness with the Father and the Son. We will see that it was necessary to undermine the doctrine of the resurrection of the Lord Jesus Christ to disorient man as to his eternal possibilities. That is, we shall see that the cover-up was less about Christ and more about man himself.

The Truth about the Resurrection Is Knowable

By way of prelude to our historical researches, let us establish that the truth is indeed knowable. Man, if sincere, cannot fail in his pursuit

---

[4] See a survey of these issues in Paul Y. Hoskisson, ed., *Historicity and the Latter-day Saint Scriptures,* (Religious Studies Center: Brigham Young University, 2001).

of unseen spiritual truth, for three reasons: First, because Truth *is*; it exists and, although it shines into the dark world (see D&C 88:7), it is largely unperceived (88:49). The Light of Truth forms the matrix of reality, all creation being permeated with and upheld by it (88:12–13). Man's very life is possible only because of this Light: "I am the true light that is in you, and . . . you are in me; otherwise ye could not abound" (88:50). Man lives and moves and has his being in the ocean of truth.

Second, God Himself has assumed the responsibility of providing man with the means of access to the truth. One way He does that is through written scripture, which supplies a link between the material world and the unseen world of spiritual realities. The Lord has provided that when a person comes to the scripture *for truth* (contrary to Korihor's contention in Alma 30:15), that person can indeed know things beyond the sensory or scientifically perceived, by means of the spirit of prophecy and revelation that always attends the scripture: "The Book of Mormon and the holy scriptures are given of me for your instruction; and the power of my Spirit quickeneth all things" (D&C 33:16).

Scripture not only provides man with information but also opens a revelatory conduit between God and man, providing unlimited access to truth. The Lord told the brother of Jared that "if he would believe in him . . . he could show unto him all things" (Ether 3:26). This revelatory relationship is the primary purpose for the coming forth of scripture. Scripture, then, is a valuable tool for uncovering truth because by its very nature it provides its *own* verification while leading to additional revelation. God has promised that He will confirm all His words to the believer (Mormon 9:25). We see also that the Lord not only requires man to believe revealed scripture but will judge him by what is written (see 2 Nephi 25:22).

Because of this revelatory power, scripture does not have to be definitive. In fact, scripture cannot possibly be definitive, words being a poor vehicle for the fulness of truth. The informed reader will sense that scripture contains only the key words of the unwritten fulness of the gospel. These key words are designed to provoke the reader both to activity and inquiry. Therefore, the words in the scripture do not have to be perfect or complete or even empirically verifiable; they need only indicate the means and direction of inquiry. A person can ask God if Jesus was truly resurrected, and he can know. Then, the seeker learns that the instruments for finding truth have more to do with desire than with science.

The truth about the resurrection is not optional. God, having provided access to all truth, places on man an obligation to seek it. Jesus told the Nephites that it was only the lesser part of the things He taught that would be distributed among them and later generations as a test of their true desire: "And when they shall have received this, which is expedient that they should have first, to try their faith, and if . . . they shall believe these things then shall the greater things be made manifest unto them. And if . . . they will not . . . , then shall the greater things be withheld from them, unto their condemnation" (3 Nephi 26:9–10). We must read and ask questions that provoke the revelation of the greater things.

The third reason that man cannot fail in his pursuit of truth is that man has a truth-discerning nature. The student of truth learns that it is his own primeval nature that makes scripture work for him as a revelatory tool. Man is able to recognize truth, not only because he lives in the ocean of truth but also because he himself is created of the stuff of truth—namely, the spirit of truth, of light, and of intelligence (see D&C 93:23–29). He is created out of the supernal substance known as *holy spirit*. We see that men and women are "instructed," even *structured*, in such a way that they know, or may know, truth from error (see 2 Nephi 2:5). In short, the children of God are quintessentially material truth. Their condemnation rests in their being made of truth but rejecting it, by which rejection they defile themselves as temples of truth (see D&C 93:31, 35). Truth is so much man's native physiology that when the "seed" of truth is presented to him, he must *resist* it by his unbelief in order not to recognize it (see Alma 32:28). Brigham Young taught that God has placed in His children "a system of intelligence that attracts knowledge, as light cleaves to light, intelligence to intelligence, and truth to truth. It is this which lays in man a proper foundation for all education."[5] As offspring of a God, man has the seeds of Deity within him.

So the issue before us is that the pursuit of the truth of the resurrection is not so much an empirical problem as a spiritual one. In fact, the avowed pursuit of truth is often a sham. It is not the truth that is sought at all but rather vindication for a personal objective.

The Plots against the Body of Man and the Body of God

Now we begin our investigations in the early Christian period. The Apostle John, alert to the dynamics of deception and foreseeing that the Lord's resurrection would become a target of opposition, warned the early Saints, "Every spirit that confesseth not that Jesus Christ is come in the flesh is not of God: and this is that spirit of antichrist, whereof ye have heard that it should come; and even now already is it in the world" (1 John 4:3). John's words reveal that an early plot was already afoot to refute the Lord's corporeality and thereby invalidate the testimonies of that moment in real time when Jesus Christ received His resurrected, glorified body.

To dismantle a doctrine as compelling as the resurrection, it is necessary to find the right avenue to man's mind. Deception can be presented in spiritually irresistible guises. For example, implicit in man's nature is the desire for holiness; therefore, when the serpent whispered to Eve, "Your eyes shall be opened, and ye shall be as gods, knowing good and evil," and when Eve saw that the tree was "to be desired to make her wise" (Moses 4:11–12), she ate the fruit. The desire for that higher knowledge that leads to greater holiness not only opens the way to exaltation but also drives many an apostate movement.

And so it was that even before the Lord's advent in the meridian of time, there lay in wait among both Jews and Greeks a philosophy called *asceticism*. Asceticism holds that the material world is devoid of spiritual value. Some of its proponents, particularly the Gnostics

---

[5] JD 1:70–71.

(whom we will discuss later), believed that the world was, in fact, created by the accident of a demigod and that only by renouncing this world and rejecting various functions of the body (such as marriage relations and the procreation of children) could a person attain the highest spiritual state. Asceticism promoted rigorous abstention from physical gratification. The practice was based on the belief that renunciation of the desires of the flesh (often including self-mortification) can bring man to the highest holiness.

It was a short step from despising the human body to the position that God Himself, the highest manifestation of holiness, had to be immaterial. Therefore, ascetic renunciation of the flesh and rejection of an anthropomorphic God developed concomitantly in Christianity; both represented a clear break with Old Testament religion and with the tenor of Christ's and the Apostles' teachings on the Father and the Son and on the physical body, with its associated functions of marriage and reproduction. As man became increasingly ascetic in his practice, his notion of God took on ascetic features as well. Ascetic man had recreated God in his own image.

Perceiving that man's eternal nature and destiny as well as his likeness to God would be the prime doctrinal targets of apostasy, the Prophet Joseph Smith remarked:

> If men do not comprehend the character of God they do not comprehend themselves. . . . It is necessary for us to have an understanding of God at the beginning; if we get a good start first we can go right, but if you start wrong you may go wrong. But few understand the character of God. They do not know [that] they do not understand their relationship to God. . . . Hear it, O Earth! God who sits in yonder heavens is a man like yourselves. That God, if you were to see him today, that holds the worlds, you would see him like a man in form like yourselves. . . . And you have got to learn how to make yourselves God, king and

priest, by going from a small capacity to a great capacity to the resurrection of the dead to dwell in everlasting burnings.[6]

The Prophet's words provide an essential capsule of the doctrine of Christ: man exists on a continuum with God. But asceticism acted catalytically on the text, doctrine, and practice of the earliest Christian church and transformed the New Testament teachings concerning God, Christ, and man.[7]

Christian asceticism had a hellenistic origin, mediated through Judaism. An example of an influential Jewish philosopher and scholar is Philo of Alexandria, who, for his cosmic view, depended heavily on Plato. Philo wrote during the early first century A.D., reflecting the great questions of the age: "Where was my body before birth, and whither will it go when [I have] departed? . . . Whence came the soul, whither will it go, how long will it be our mate and comrade? Can we tell its essential nature? When did we get it? Before birth? But then there was no 'ourselves.' What of it after death? But then we who are here joined to the body, creatures of composition and quality, shall be no more, but shall go forward to our rebirth, to be with the unbodied, without composition and without quality."[8]

Here we see the spirit of inquiry gone awry. Philo started with good questions, but he came to wrong conclusions about a bodiless heaven; he did this by mingling the philosophies of men with the word of God. The literature of the early Christian apologists and the Gnostic literature of the first two centuries possess many allusions to

---

[6] Andrew F. Ehat and Lyndon W. Cook, eds., *The Words of Joseph Smith: The Contemporary Accounts of the Nauvoo Discourses of the Prophet Joseph* (Provo, Utah: Religious Studies Center, Brigham Young University, 1980), 340, 343–45.

[7] For more detailed information on this ascetical revolution and its precipitation of the great apostasy, see the author's "The Influence of Asceticism on the Rise of Christian Text, Doctrine, and Practice in the First Two Centuries," unpublished dissertation, BYU, 1989.

[8] Riemer Roukema, *Gnosis and Faith in Early Christianity*, trans. J. Bowden (Harrisburg, Pennsylvania: Trinity Press, 1999), 60–61.

the writings, or at least the thought-world, of Philo and hellenized Judaism.

But asceticism also gained access to Christianity through men trained in Greek philosophy who, when they were converted to Christianity, brought their philosophical constructs with them. Some of these earliest Christian converts wrote to explain and defend Christianity to their Greek-educated friends but tried to couch the Christian message in terms their intellectual friends could accept. Because the basic truths of the gospel (such as the efficacious suffering of a physical God and His miraculous resurrection) cannot be reshaped in Greek philosophical terms, these early Christians succeeded not in clarifying and preserving the gospel but in distorting it. It soon became unsophisticated to accept the plain truths taught by the Lord Jesus Christ. As a result of such influences, the once-anthropomorphic Father and Son became amorphous, passionless, and ascetic. The Savior was born into this philosophical climate where the suppression of the value of the body of both man and God had already been at least partially successful.

Ascetical ideas permeated the early church. What Paul called the "mystery of iniquity" was at work early on (see 2 Thessalonians 2:7). He wrote to the Saints to defend the Lord's physical resurrection, stressing that "in him dwelleth all the fulness of the Godhead *bodily*" (Colossians 2:9; emphasis added). The Apostles bore fervent testimony as they and other eyewitnesses fanned out around the Mediterranean: "We have seen and we know."

But the Apostles were fighting a losing battle, and soon after they were gone, another group called the Apostolic Fathers took up the cause against the wave of unbelief and doctrinal distortion. These were Christian leaders who had had some association with the Apostles and who wrote letters or treatises to strengthen their congregations. One of these, Justin Martyr (c. A.D. 100–165), wrote against ascetical ideas of the resurrection: "They who maintain the wrong opinion say that there is no resurrection of the flesh . . . and they abuse the flesh, adducing its infirmities, and declare that it only is the cause of our sins, so that if the flesh, say they, rise again, our infirmities also rise with it. . . . By these and such arguments, they to distract men from the faith. . . . These persons seek to rob the flesh of the promise."[9]

Another father, Ignatius of Antioch (early second century), a bishop who suffered martyrdom in Rome, wrote by way of testimony and also by warning, concerning resurrection and the "apostate dragon":

> Ignatius . . . rejoices in the Passion of our Lord without doubting, and is fully assured in all mercy in his resurrection. . . . Therefore as children of the light of truth flee from division and wrong doctrine. [10]

> If any one confesses the truths mentioned, but calls lawful wedlock and the procreation of children, destruction and pollution . . . such an one has the apostate dragon dwelling within him. If any one confesses the Father, and the Son and the Holy Ghost, and praises the creation, but calls the incarnation merely an appearance [the heresy of *doceticism*], and is ashamed of the passion, such an one has denied the faith, not less than the Jews who killed Christ.[11]

The implication in their letters is that if you don't believe in the literal, physical resurrection of Jesus Christ, you are not a Christian.

The Doctrine of the Body

The "apostate dragon" had his own agenda and scrambled the thinking of early Christian authors on four main topics: the human body; marriage (and women); procreation; and God's body, including the body of the resurrected Son. The devaluing of the body developed along with a sense of world-alienation in hellenistic Judaism, which

---

[9] *Fragments of the Lost Work of Justin on the Resurrection, II,* translated in Alexander Roberts and James Donaldson, eds., *Ante-Nicene Fathers,* 10 vols. (Grand Rapids, Michigan: Eerdmans, 1989), 295.

[10] *Philadelphians,* Greeting, 2.1.

[11] *Philadelphians,* long recension, 6.

459

Christianity absorbed remarkably soon after its inception. It caused some Christian theologians to teach that a better world would arrive if people would quit perpetuating the present fallen order by their acts of reproduction—that is, it would be better to let the material world and fallen man die out.[12] The higher spiritual order would come on earth only when the old one, the one Eve had precipitated, had ceased. This alienation from bodily reality led to the assumption that God is unlike man, having no relationship to the material world; God is something other.

I include here for our discussion some teachings from the latter-day Brethren who enlarge on three purposes for the union of flesh and spirit in the physical world, for in man's body are the secrets of his destiny.

As mentioned above, man was created before the foundations of the earth out of material that we call *holy spirit*, a highly capacitated, holy, and pure substance. Therefore, in the premortal world, man was already a highly developed spirit being. As Parley P. Pratt taught, man's spirit substance was more elastic, subtle, and refined than the fleshy body he would take, but seeing that a tabernacle would give him power to develop and exalt himself in the scale of intelligence, both in time and eternity,[13] he stooped to take a body.

President John Taylor explained, "It was by the union of their spirits, which came forth from the Father as the 'Father of Spirits,' with earthly bodies, that perfect beings were formed, capable of continued increase and eternal exaltation; that the spirit, quick, subtle, refined, lively, animate, energetic, and eternal, might have a body through which to operate . . . [in order that the spirits not be left to] spend their force at random, or remain dormant, or useless, without those more tangible, material objects, through which to exercise their force. Thus, then, was the body formed as an agent for the spirit."[14]

---

[12] Augustine, *De Bono Conjugali*, 17–20; *De Bono Viduit*, 9–11, 23–28.

[13] JD 1:7–9.

[14] John Taylor, *The Government of God* (London: S.W. Richards, 1852), 77–78.

Therefore, even though the body was made of grosser materials than the spirit, it "was necessary as an habitation for it that, it might be clothed with a body, perfect in its organization, beautiful in its structure, symmetrical in its proportions, and in every way fit for an eternal intelligent being; that through it, it might speak, act, enjoy, and develop its power, its intelligence and perpetuate its species. . . . They [the spirits] had the intelligence before, but now they saw a way through which to develop it."[15]

The spirits needed not only a body but also, to attain to their high destiny, experience in the two major spheres of which the universe is constituted. Elder John A. Widtsoe explained, "The universe is dual: spiritual and material, composed of 'spirit-element' and 'matter-element.' These two realms are closely interwoven, perhaps of the same ultimate source; yet they are distinct in their nature. Mastery of the universe means acquaintanceship with, and control of both of these elemental divisions of the universe in which we live."[16]

In addition, as part of the eternal progress of spirit man, the body provided protection against the powers of darkness. That is, as Brigham Young taught, "the body was formed expressly to hold its spirit and shield it."[17] And as Joseph Smith taught, "There are things which pertain to the glory of God and heirship of God with Christ which are not written in the Bible. Spirits of the eternal world are diverse from each other as here in their dispositions. . . . As man is liable to enemies there [the other world] as well as here it is necessary for him to be placed beyond their power in order to be saved. This is done by our taking bodies (keeping our first estate) and having the power of the resurrection pass upon us whereby we are enabled to gain the ascendancy over the disembodied spirits. The mortification of Satan consists in his not being permitted to take a body."[18]

[15] Taylor, The Government of God, 78–79.
[16] John A. Widtsoe, Evidences and Reconciliations (Salt Lake City: Bookcraft, 1960), 72.
[17] Brigham Young, in JD, 9:140.
[18] Ehat and Cook, Words of Joseph Smith, 208.

Clearly it is body-envy that drives the hosts of the dark side and explains a good deal about the plot against the Lord's resurrection. We recall that Adam was given dominion, that is, priesthood power,[19] over all the earth, that through his and his posterity's dominion the earth might be governed by the power of the Spirit and the priesthood. But the adversary, the great usurper, seized his opportunity to wrest the dominion of this earth and set about to take possession of the bodies of men and, through their instrumentation, rule the earth, making it and its fulness his servants.[20] By gaining control of the bodies of men through keeping them ignorant, Satan could overcome here and hereafter man's power to thwart him in his empire-building. Creating confusion about the Lord's resurrection while occupying the minds and bodies of men promoted the adversary's purposes for extending his own dominions.

The adversary's success created a striking contrast between the church of the Apostles and the later one influenced by asceticism. Before asceticism, the human body was viewed as the vehicle and channel of the Spirit of God, the means by which men and women give and receive love and exercise their functions as incarnate children of God. The New Testament teaches that the body is the temple and is therefore a bridge between heaven and earth. But after asceticism, as one scholar observes, "we find ourselves . . . with an idea that bodily needs and pleasures are unspiritual and to be shunned (what terrible consequences this has had for women, who by their very nature represent bodily needs and pleasures to the all-male body who developed Christian theology!)."[21]

Plots directed against the resurrection as part of the Lord's atonement also surfaced in various guises during this period. From within the ascetical environment of the second century A.D., a complex

---

[19] Joseph Smith, *Teachings of the Prophet Joseph Smith*, selected by Joseph Fielding Smith (Salt Lake City: Deseret Book, 1976), 157.

[20] Erastus Snow, JD, 19:275–76.

[21] Margaret Barker, *The Lost Prophet: The Book of Enoch and Its Influence on Christianity* (London: SPCK, 1988), 74–75.

religious movement called Gnosticism spread among the Christians. Gnosticism is based on the Greek word γνῶσις (*gnōsis*) ("knowledge"), which was supposedly revealed knowledge about God and the origin and destiny of mankind. It was the body of secrets by which the spiritual element in man, as opposed to the physical, could receive redemption.

Because the Lord had indeed instructed a select few in the higher doctrines of the kingdom, there was heightened interest among Christians in the secrets pertaining to holiness and entrance into the kingdom of God. Many of these gnostic "secrets" pertained to ascetical practices and found their way into Christian practice.

But specifically, with respect to the Atonement and the resurrection, there was allied with Gnosticism the belief called Docetism (from Greek δοκέω (*dokeō*), "to seem") which considered the humanity and sufferings of the earthly Christ as *apparent* rather than real. God could not be material; therefore, He could not suffer. We can tell that Gnosticism made its appearance early in the Church's history since it is frequently addressed in the New Testament (1 John 4:1–3; 2 John 1:7; Colossians 2:8–9; 1 Timothy 6:20, footnote, "Keep that which is committed to thy trust, avoiding . . . disputations of what is falsely called knowledge [gnosis]").

Docetic Christology held that Christ did not suffer on the cross but was represented at the crucifixion through a substitute, whether the earthly and bodily Jesus or another man, such as Judas Iscariot or Simon of Cyrene, who changed places with Him just *before* the crucifixion. The Docetists taught that Christ only *seemed* to suffer but in reality did not. A docetic text, the *Apocalypse of James*, has Christ say to the grieving James, "Never have I experienced any kind of suffering, nor was I (ever) tormented, and this people nowhere did any evil . . . I suffered (only) according to their view and conception."[22] As the text continues, it implies that it is acceptable that the body of this

---

[22] Kurt Rudolph, *Gnosis: The Nature and History of Gnosticism* (Harper & Row: San Francisco, 1983), 167–68.

earthly Jesus was destroyed because it had an unholy origin, recalling the old gnostic idea that the material creation of the earth was an unfortunate accident.

A famous image in Gnosticism is the "laughing Messiah," which appears in another apocryphal work, the *Apocalypse of Peter*, where the Lord says to Peter (who is supposedly beholding a vision of the crucifixion), "He whom you see on the . . . cross, glad and laughing, this is the living Jesus. But he in whose hands and feet they drive nails is his fleshly [likeness], it is the substitute . . . whom . . . they put to shame, the one who originated after his likeness."[23] In Gnosticism, the resurrection takes place *before* or at the same time as the crucifixion.[24]

Though the Gnostics were considered heretics by "mainstream" Christianity during the first few centuries, later fathers harvested the confusion that Gnosticism had sown. Hugh Nibley comments ironically on various interpretations of the resurrection among later writers:

> The early apologist Athenagoras insisted that life would be utterly wasted without the resurrection; it is the resurrection which gives everything in human life its meaning. Yet Rufinus tells us that "after the resurrection, all will be spirit—no bodies." But, says Hilary, there must be a physical resurrection. The scriptures say it's so. But it can only be for the wicked. Only they deserve that kind of punishment. That's certainly a desperate twist. Gregory of Nyssa, one of the four great Greek Fathers, said if you must "gape after sensual enjoyment, and ask . . . 'Shall we have teeth and other members [after the resurrection?],' . . . the answer is yes, since the scriptures [won't allow us to deny it—they] are perfectly clear, we shall have all our members—but we will not make use of them." Jerome himself says yes, our bodies will be resurrected, but since we have no further need of bodies, the minute we are resurrected, we will start to dissolve; and "all matter will return

---

[23] Rudolph, 170.
[24] Rudolph, 171.

to the *nothing* (*nihilum*) from which it was once made"—back to Nirvana. But is that satisfaction? I ask.[25]

With such thinking, the field had been sown with ascetical ideas in preparation for the Savior's advent. These ideas were designed to cause people to reject the truth about His resurrection. They were calculated to disparage the body and the material world and promote a bodiless God in an immaterial heaven. As these ideas flourished, many later Christians could not conceive of a glorified, resurrected Savior whose redeeming act would secure the resurrection of the entire material world to sanctify it for an eternal, glorified existence. The diabolical aim was to invalidate the miracle of the resurrection so that man would miss his mark.

The Plot against Man's Oneness with the Father and the Son

One way to distort doctrine is to define it sufficiently narrowly that the fulness of its meaning is obscured. Monotheism was such a doctrine among the Jews. Traditional monotheism is captured in the passage "Hear, O Israel: The Lord our God is one Lord" (Deuteronomy 6:4), or, as an alternate translation, "The Lord our God is one."

This passage traditionally refers to the belief that there is only one true God as opposed to the many gods of the ancient Near East. The Jews clung tenaciously to this definition of monotheism as one of the chief elements of their religion that set them apart from their pagan neighbors and gave them power with Jehovah, the true God. The problem with this traditional understanding of monotheism is that it can preclude multiple gods, such as a Son of God, as well as the possibility of there being additional gods—that is, it can be used to

---

[25] Hugh Nibley, *Temple and Cosmos: Beyond This Ignorant Present*, The Collected Works of Hugh Nibley, Volume 12, ed. Don E. Norton (Salt Lake City: Deseret Book and FARMS, 1992), 358.

preclude not only the separate, resurrected body of the Son but also the deification of resurrected man.

Various groups among the Jews had to tamper with scripture to reduce monotheism to this one-dimensional definition. Jesus, referring to their history of scriptural revision, reproached the scribes and interpreters of the law around him: "Woe unto you, lawyers! for ye have taken away the key of knowledge, the fulness of the scriptures; ye enter not in yourselves into the kingdom; and those who were entering in, ye hindered" (JST Luke 11:52).

A word about revision of scripture: if one is going to tamper with scripture, he must first get rid of the concept of continuing revelation in order to make his subjects dependent on written sources rather than on revelatory ones. That is, one way to foster dependency is by providing a definitive, written law and tradition, not as adjunct to, but as substitute for, continuing revelation.

In fact, it seems that in every dispensation, theologians elect themselves to make text and tradition transcend continuing revelation. The attempted coup results in the suppression of truth about the penetrable veil. Practicing religion their own way, these priests and scribes obtain power over others who might otherwise penetrate the veil for themselves, gain access to personal revelation, and become a threat to the established power base. One who has opened his own conduit to heaven will think for himself, will insist on the truth, and will have recourse to God rather than to the priestly group for his needs. Therefore, the priests must adjust scripture, making revisions here and redactions there, to make God as unobtrusive as possible, de-anthropomorphizing him, misrepresenting His character, and distancing him to the most remote part of heaven—or better, to oblivion. "They . . . teach with their learning, and deny the Holy Ghost, which giveth utterance. And they deny the power of God, the Holy One of Israel. . . . For this day he is not a God of miracles; he hath done his work" (2 Nephi 28:4–6).

In an interesting interchange on this subject between Jesus and His disciples, He instructed them to teach the people to ask of God, to seek, to knock—that is, to seek behind the veil. The disciples protested, fearing that the people would say, "We have the law for our salvation, and that is sufficient for us." The Savior replied, "Thus shall ye say unto them, What man among you, having a son, and he shall be standing out, and shall say, Father, open thy house that I may come in and sup with thee, will not say, Come in, my son; for mine is thine, and thine is mine?" (JST Matthew 7:12–17). The cover-up here has to do with blocking man's awareness that a benevolent God seeks to interact with him from a world of miracles to make him a co-heir in the kingdom of heaven.

Let us consider two such groups of revisionists here, with reference to the resurrection, in reverse chronological order: Jewish scribes/rabbis and the so-called Deuteronomists. At least by intertestamental times, the Soferim (scribes and successors to Ezra, c. 458 B.C.) and the Tannaim (rabbis who passed on the oral tradition in the later intertestamental and early Christian period) found the biblical anthropomorphisms offensive and took steps to make small emendations, which they described as "biblical modifications of expression."[26] These scribes are those to whom the Savior referred above. They effectively laid the foundation for the rejection of the Messiah when He should appear by preparing the people to reject an anthropomorphic God—or His Son.

The second group, the Deuteronomists, are controversial in current biblical scholarship with respect to who they were, when they flourished, and what they did. In this regard, one scholar remarked, "The Deuteronomists have sometimes been praised or blamed for virtually every significant development within ancient Israel's religious practice."[27] Nevertheless, I will give the reader a flavor of the praise and blame this group may merit as it worked on parts of our Old

---

[26] Cecil Roth, in *Encyclopaedia Judaica* (Jerusalem: Keter Publishing House, 1972), s.v. "Anthropomorphism," 2:53.

[27] R. J. Coggins cited by Linda S. Schearing in *Those Elusive Deuteronomists, The Phenomenon of Pan-Deuteronomism*, ed. Linda S. Schearing and Steven L. McKenzie (Sheffield, England: Sheffield Press, 1999), 13.

Testament, since with the perspective of such restored scriptures as 1 Nephi 13, we know that the Bible was well worked over.

During, or perhaps following, the exile of the Jews to Babylonia, these Deuteronomists apparently made revisions to scripture, suppressing several ancient truths. One scholar describes this group as "the great theologians, freeing Israel from the vagaries of inspiration to the stability and reason of books of law."[28]

Another scholar, seeing these revisionists not as liberators but as censors of truth, remarked, "The Deuteronomists suppressed the anthropomorphism of the older tradition and any idea of the visible presence of God was abandoned. . . . The old concept of a human form present in the temple was no longer tenable, and the ancient descriptions of theophanies derived from temple ceremonial were no longer acceptable. The Deuteronomists rewrote the tradition: 'Then Yahweh spoke to you out of the midst of the fire; you heard the sound of the words but saw no form; there was only a voice'" (Deuteronomy 4:12).[29]

Their strict monotheism promoted rejection not only of all the hosts of heaven (which would also preclude the ministry of angels) but also a Son or sons of God. The thrust was to do away with the world of miracles—to which the resurrection belongs. "The Deuteronomists were fervent monotheists, which has led us to believe that all the Old Testament describes a strictly monotheistic religion. They also said that God could not be seen, only heard. There were, however, ancient traditions that said otherwise in each case; there was . . . a belief in *a second divine being who could have human form* and this became the

---

[28] Reference to William Doorly's *Obsession with Justice: The Story of the Deuteronomists* (New York: Paulist, 1994), cited by Kevin Christensen, *Paradigms Regained: A Survey of Margaret Barker's Scholarship and Its Significance for Mormon Studies* (Provo, Utah: FARMS, 2001), 8.

[29] Margaret Barker, *The Great Angel: A Study of Israel's Second God* (Louisville: Westminster/John Knox Press, 1991), 99–100; cited by Christensen, *Paradigms Regained*, 14.

basis of Christianity."[30] With a half-truth, they tried to suppress the more comprehensive understanding of monotheism.

The Apostle Paul testified that "there is but one God, the Father . . . and one Lord Jesus Christ" (1 Corinthians 8:6)—that is to say, two divine beings, which seems to be a contradiction of Deuteronomy 6:4. However, as one scholar observes with significant insight, "if . . . [the Apostle's statement] was a statement of the unity of Yahweh as the one inclusive summing up of all the heavenly powers, the 'elohim, then it would have been compatible with belief in God Most High also."[31] Here she touches on the possibility of additional dimensions in the doctrine—that is, that monotheism may have to do not with only one among many but with the union of all the heavenly host under the Most High God, the Father of all. This insight allows us to find enlarged meaning in the alternative translation given above, "The Lord our God is *one*" (Deuteronomy 6:4; emphasis added).

Our doctrine teaches that there is a Father and there is a Son who are separate beings; but also, that the Son encompasses both the Father and the Holy Spirit within Himself (see Mosiah 15:1–5, 7; Alma 11:28–29, 38–39, 44). These two seemingly contradictory ideas lead us to explore a more comprehensive definition of the term *monotheism*.

The Power to Be One

One approach to the question of how multiple Gods can be one begins with a definition of God and His power to make one of many. It would appear that *holy spirit* (mentioned above) constitutes the origin of both man and God. Brigham Young taught that the attributes this holy spirit possesses "can be made manifest only through an organized personage."[32] That is to say, a God is the highest manifestation of organized holy spirit.

[30] Margaret Barker, *The Gate of Heaven: The History and Symbolism of the Temple* (London: SPCK, 1991) 7; Christensen, *Paradigms Regained*, 24; emphasis in original.
[31] Barker, *The Great Angel*, 192–93; cited by Christensen, *Paradigms Regained*, 26.
[32] *JD*, 10:192.

Charles W. Penrose further explained, speaking of the organization of holy spirit into a God:

> The perfection of its [holy spirit's] manifestation is in the personality of a being called God. That is a person who has passed through all the gradations of being, and who contains within Himself the fullness, manifested and expressed, of this divine spirit. . . . If you see a man you behold its most perfect earthly manifestation. And if you see a glorified man, a man who has passed through the various grades of being, who has overcome all things, who has been raised from the dead, who has been quickened by this spirit in its fullness, there you see manifested, in its perfection, this eternal, beginningless, endless spirit of intelligence. Such a Being is our Father and our God, and we are following in His footsteps. . . . He is a perfect manifestation, expression and revelation of this eternal essence, this spirit of eternal, everlasting intelligence or light of truth.[33]

Even though this perfect being, organized of a fulness of holy spirit, retains both his body and his personal identity to all eternity (having eternal fundamental parts),[34] this person nevertheless has the capacity to enter into full union with an infinite number of other glorified persons.

Elder Orson Pratt explained what it means for two or more persons to come into at-one-ment:

> Jesus could with all propriety say, when speaking of the knowledge he had, "The Father is in me, and I in him . . . and inasmuch as you have received me, I am in you, and you in me." That is as much as to

---

[33] JD, 26:24–25.

[34] "There is no fundamental principle belonging to a human system that ever goes into another in this world or in the world to come; I care not what the theories of men are. We have the testimony that God will raise us up, and he has the power to do it. If any one supposes that any part of our bodies, that is, the fundamental parts thereof, ever goes into another body, he is mistaken" (Joseph Smith, *History of the Church*, 8 vols. [Salt Lake City: Deseret Book, 1967], 5:339).

say, that "not the whole of me is in you, because you are imperfect: but inasmuch as you have received the truth I have imparted, so much of me is in you, for I am the truth, and so much of you dwells in me." And if you should happen to get a knowledge of all the truth that he possesses, you would then have all of his light, and the whole of Christ would then dwell in you. . . . Hence we see that wherever a great amount of this intelligent Spirit exists, there is a great amount or proportion of God, which may grow and increase until there is a fulness of this spirit, and then there is a fulness of God.[35]

The *Lectures on Faith* offer further clarification on the nature of the relationship of oneness:

The Father and Son possess the same mind, the same wisdom, glory, power, and fulness; filling all in all—the Son being filled with the fulness of the mind, glory, and power, or, in other words, the spirit, glory, and power of the Father—possessing all knowledge and glory, and the same kingdom; sitting at the right hand of power, in the express image and likeness of the Father—a Mediator for man—being filled with the fulness of the mind of the Father, or, in other words, the *spirit* of the Father; which spirit is shed forth upon all who believe on his name and keep his commandments; and all those who keep his commandments shall grow up from grace to grace, and become heirs of the heavenly kingdom, and joint heirs with Jesus Christ; possessing the same mind, being transformed into the same image or likeness, even the express image of him who fills all in all; being filled with the fulness of his glory; and *become one* in him, even as the Father, Son, and Holy Spirit are one."[36]

The foregoing is an illuminating description of the society of the Gods and the state called exaltation. Their unity, possible only

---

[35] JD, 2:342–43.

[36] *Lectures on Faith* (American Fork, Utah: Covenant, 2000), Lecture 5; emphasis added.

through adherence to eternal law, consists in the voluntary merging of will, purpose, and consciousness. This perfect communion among such beings provides the glorified environment of exaltation. That is, without this perfect union, not only of body and spirit but also with other beings, there is no fulness of joy and no exaltation.

The principle of joy in spiritual union operates in every sphere. Even on earth the union of two people in righteousness can approach a state of ecstasy. And a community of such people on earth constitutes that Zion whose citizens the scriptures describe as "partakers of the heavenly gift" (4 Nephi 1:3) and "surely there could not be a happier people" (4 Nephi 1:16).

We understand, then, that individually organized spirits are like permeable vessels in that they can be filled to capacity with a fulness of holy spirit that unites them perfectly with all other beings who have also been so filled. Being filled with holy spirit, they possess also a fulness of all truth, for "the spirit knoweth all things" (Alma 7:13). They therefore enter into a state of congruence, and in this state they share a perfect consciousness of things as they are, as they were, and as they are to come (D&C 93:24; Jacob 4:13). Jesus declared, "I am in the Father, and the Father in me, and the Father and I are one—[I am] the Father because he gave me of his fulness" (D&C 93:3; compare JST Luke 10:23: "All things are delivered to me of my Father; and no man knoweth *that the Son is the Father, and the Father is the Son, but him to whom the Son will reveal it*" [emphasis added]). It is left to the seeker to receive the revelation that the Savior refers to above.

Thus, when Abinadi declared that the Father and the Son are "one God, yea, the very Eternal Father of heaven and of earth" (Mosiah 15:4) and when Amulek answered that the Son of God "*is* the very Eternal Father of heaven and of earth" and also that "Christ the Son, and God the Father, and the Holy Spirit . . . *is* one Eternal God" (Alma 11:39, 44; emphasis added), they described true monotheism.

We might then discover that monotheism really refers to the fact that *there is only one God who has the power to bring an infinite number of beings into perfect union with Himself in a fulness of joy.* Monotheism both implies exaltation and defines the nature of exaltation. This vision of people living in the joy of conscious union, in a society coupled with eternal glory (D&C 130:2), is far different from that of a lone God, existing in isolation from an inferior creation that can never be one with Him.

An additional implication in this broader definition of monotheism is that it also defines the order of the heavens—that is, all the hosts of heaven live and move and have their being in the administration of one will and one language. As we read, the angels speak by the power of the Holy Ghost (2 Nephi 32:3), which is the medium of the Father's will. We read also that a person can enter into this heavenly order by receiving the Holy Ghost, that he might be shown all things that he should do (32:5); such a person, to a degree at least, is at one with the order of heaven even as he walks the earth.

Yet another aspect of this enlarged definition of monotheism implies that a resurrected person may, if faithful, receive the priesthood keys to assist in the work of the resurrection of the dead. Brigham Young taught: "The keys of the resurrection . . . will be given to those who have passed off this stage of action and have received their bodies again. . . . They will be ordained, by those who hold the keys of the resurrection, to go forth and resurrect the Saints, just as we receive the ordinance of baptism, then the keys of authority to baptize others for the remission of their sins. . . . [In addition,] when our spirits receive our bodies, and through our faithfulness we are worthy to be crowned, we will then receive authority to produce both spirit and body . . . [and] be ordained to organize matter."[37]

Returning to our historical narrative, I repeat that all these beautiful doctrines had been obscured even before the Savior's birth, and most references to the resurrection, man's destiny, and the power of Christ had been deliberately removed from the Old Testament.

Therefore, the threat of the coming to light of these doctrines often provoked a mean spirit among those who had thought them

---

[37] JD, 15:137.

done away. The Jews' attachment to traditional monotheism and their previous attempts to de-materialize God led them to respond violently to the Lord when He said to them, "Before Abraham was, I am" (John 8:58; "I am" signifies the Greek equivalent of the Hebrew tetragrammaton, *Yahweh*, or Jehovah, which was forbidden to be spoken). They took up stones against Him, as they did when He declared, "I and my Father are one." When the Lord asked them why, the Jews responded, "Because that thou, being a *man*, makest thyself God" (John 10:30, 33; emphasis added). To the Jews, there could not be multiple gods, and God could not be in the form of a man.

If God was only a nebulous substance or bodiless spirit, that is, something entirely other than man, then man could not become like Him. It was concluded, then, in those early centuries, that man did not exist on a continuum with God. This conclusion is the crux of the cover-up.

As a result, while debates over monotheism and the nature of God and Christ combined with power politics, the Church fathers felt obligated to merge the Father and the Son, making them consubstantial. This step effectively abolished Christ's mediating role, which was subsequently filled by holy relics, holy virgins, holy martyrs, and holy celibates. Therefore, when the Christian bishops from all over the Mediterranean convened at Nicaea (A.D. 325) to draft a creed on the nature of the Father and Son, the first official council of the Christian Church sanctioned a new form of monotheism: the non-material, three-in-one God. In consequence of the declining importance of the Atonement and of Christ as mediator, the felt need for mediation, as in the administration of penance, was filled by ascetic men. In the act of mediation, the holy man accrued to himself considerable authority in the dispatch of his office. Thus, these mediators established their power base of holy intermediaries on an anti-Christ creed.

An example of a descendant of the Nicene Creed is the Thirty-nine Articles of the Church of England. This creed, typical of orthodox Christian creeds, contains a familiar phrase: "There is but one living and true God, everlasting, without body, parts, or passions; . . . and in unity of the Godhead there be three Persons, of one substance, power, and eternity: the Father, the Son, and the Holy Ghost."[38] Thus, there trickled down through the centuries to the present day an effete counterfeit of the true God.

For the most part, Gnosticism triumphed. The doctrinal distortions perpetrated in the early Christian period laid the foundation of modern Christian tradition and confusion. And even while today among contemporary scholars there are those who cling to the belief that the physically resurrected Jesus was present for anyone to see, even photographable,[39] most modern theological literature yields up only the distorted progeny of those confusions spawned in the early Christian period.[40]

---

[38] John H. Leith, ed., *Creeds of the Churches: A Reader in Christian Doctrine from the Bible to the Present*, 3d ed. (Atlanta: John Knox Press, 1982), 266–67.

[39] Stephen T. Davis, "'Seeing' the Risen Jesus," in *The Resurrection: An Interdisciplinary Symposium on the Resurrection of Jesus* (Oxford: Oxford University Press, 1997), 146. He asks the question that a Latter-day Saint might ask: "Why is this view [that the risen Jesus was seen as a physical being] so commonly rejected? One sometimes gets the impression from the friends of objective visions that the notion of a physically present resurrected Jesus is somehow uncouth or outre. I do not share such feelings" (142). He includes among those who recoil at the physical Jesus the influential scholar Raymond Brown, who says, "This type of question [suggesting that Christ was indeed physical] does not show any appreciation for the transformation involved in the Resurrection" (*The Virginal Conception and Bodily Resurrection of Jesus* [New York: Paulist, 1973], 91 fn.).

[40] Davis, *The Resurrection*, 131, cites Gerald O'Collins: "Most New Testament scholars would be reluctant to assert that the risen Christ became present in such a way that neutral (or even hostile) spectators could have observed him in an ordinary 'physical' fashion." Following is a brief sampling of approaches to the Resurrection reported in this symposium: The resurrection of Christ was figurative rather than literal: Jesus rose only in the mind and hearts or the lives and dreams of His followers. The Resurrection is a way of speaking about the *awareness* of His continuing presence and empowerment among the believers, an awareness that the presence of God in Jesus is a permanent presence in the midst of the believers, an expression of God's presence in all space and time (6). The witnesses to the resurrected Savior saw an impostor, a hallucination, a mass of ectoplasm, or a sort of interactive hologram (142). The witnesses' powers of perception were enhanced by God ("graced seeing"), a mental phenomenon or the "subjective vision theory," that is, a psychogenic projection (128). The stories of the empty tomb were merely elaborations of the message of the Resurrection, likely an invention of Mark (14).

As evidence of the triumph of Gnosticism, one contemporary scholar, searching among canonical and non-canonical texts for a new approach to the resurrection, observes, coming right to doceticism in her conclusion, "This matter of the resurrection . . . still lies at the heart of everything we teach. Was the resurrection, to use the former bishop of Durham's now famous phrase, just a conjuring trick with bones? Was it a case of body snatching, as the Jewish authorities apparently claimed?" No, she says; nor was it simply a resuscitation. Rather, she proposes, He was actually resurrected *before* He suffered and died.[41] She quotes as support for this approach a verse from the apocryphal *Gospel of Philip*, reflecting the docetic texts: "Those who say that the Lord died first and then rose up are in error, for he rose up *first* and then died."[42]

Other scholars, agreeing that empirical science cannot produce evidence of the resurrection, abandon themselves to the position that knowing whether the resurrection was literal is probably not all that important, seeing that the quest for the historical resurrected Jesus is futile.[43]

---

[41] Margaret Barker, "Resurrection: Reflections on a New Approach," *Resurrection*, Stanley E. Porter, Michael A. Hayes and David Tombs, eds., *Journal for the Study of the New Testament*, Supplement Series 186 (Sheffield, England: Sheffield Press, 1999), 99; emphasis added.

[42] Barker, "Resurrection," 101, citing *The Gospel of Philip*; emphasis added.

[43] Carnley, "Response," 40. "I think we have to face the question of what difference it would actually make to faith if it were to be the case that there was a psychological cause of the Easter visions of Jesus. In other words, given that we do not have the historical evidence to rule out the possibility that the appearances may have been the product of psychological processes of bereavement amongst the disciples, what is our next move? If we are unlikely ever to be able either to prove or to disprove the thesis that the appearances were psychologically induced 'subjective visions,' rather than some kind of 'objective vision,' where do we go from here?" He concludes that the risen Christ should be sought as a "religious object in present experience, rather than just engage in what . . . is a somewhat futile quest for the historical resurrected Jesus" (42).

The Quest for Jesus

The quest is not futile. Moroni said simply, "I have seen Jesus, . . . he hath talked with me face to face. . . . And . . . I would commend you to seek this Jesus" (Ether 12:39, 41). Moroni knew that the quest is not only *not* futile but that it is imperative, because man in his cosmic journey cannot even be defined without Jesus Christ, and without a conscious relationship with Him, cannot move toward his high destiny. "I am the vine," Jesus said, "ye are the branches: He that abideth in me, and I in him, the same bringeth forth much fruit: for without me ye can do nothing" (John 15:5).

In all of man's stages of progress, there was no point in which he was not entirely dependent on God's grace, not only for his very life and breath and reason and mobility (see Mosiah 2:21) but also for every step forward in his progression. Man's transition from manhood to Godhood can be accomplished only through a power superior to him, an infinite and eternal power. "For," as President Taylor explained, "in Adam all die, so in Christ only can all be made alive." He continued, "Through Him mankind are brought into communion and communication with God; through His atonement they are enabled, as He was, to vanquish death; through that atonement and the power of the Priesthood associated therewith, they become heirs of God and joint heirs with Jesus Christ, and inheritors of thrones, powers, principalities, and dominions in the eternal worlds. And instead of being subject to death, when that last enemy shall be destroyed, and death be swallowed up in victory, through that atonement they can become the fathers and mothers of lives, and be capable of perpetual and eternal progression."[44]

But seeking to thwart these very possibilities, as we have seen, the adversary exerts an invisible agency over the spirits of men by which he darkens their minds, and, as President Taylor explained:

[44] John Taylor, *Mediation and Atonement*, (Salt Lake City: Deseret Book, 1882), 139–41; republished as *Important Works in Mormon History*, Volume 4 (Orem, Utah: Grandin Book, 1992).

[Satan] uses his infernal power to confound, corrupt, destroy, and envelope the world in confusion, misery, and distress; and, although deprived personally of operating with a body, he uses his influence over the spirits of those who have bodies, to resist goodness, virtue, purity, intelligence, and fear of God, and consequently, the happiness of man; and poor erring humanity is made the dupe of his wiles. . . . But not content with the ravages he has made, the spoliation, misery, and distress; not having a tabernacle of his own, he has frequently sought to occupy that of man, in order that he might yet possess greater power, and more fully accomplish the devastation. . . .

Man's body to him, then, is of great importance, and if he only knew and appreciated his privileges, he might live above the temptation of Satan, the influence of corruption, subdue his lusts, overcome the world, and triumph, and enjoy the blessings of God in time and in eternity.[45]

The opponents of truth have sought to establish the following premises to obtain their goals:

- Man's body has no eternal value and will not persist after this life.
- Therefore, of course, God could not have a body, nor beget children.
- Therefore, there is no literal, physical resurrection.
- Therefore, there is no world of miracles, and you don't have to believe what you cannot see.
- Therefore, there is no atonement.
- Therefore, we, humanity's self-appointed leaders, will be the people's God and will get gain by obtaining power over their minds.

---

[45] Taylor, *The Government of God* (Salt Lake City: Deseret Book, 1852), 32–46. Reprinted as *Important Works in Mormon History*, Volume 3 (Orem, Utah: Grandin Book, 1992).

Even a brief look at religious history on this planet suffices to show how successful these opponents have been.

We recognize that the Lord distinguishes between perpetrators and victims of apostasy: the perpetrators are filled with the "influence of that spirit which hath so strongly riveted the creeds of the fathers, who have inherited lies, upon the hearts of the children, and filled the world with confusion" (D&C 123:7). And the victims are found "among all sects, parties, and denominations, who are blinded by the subtle craftiness of men . . . and who are only kept from the truth because they know not where to find it" (123:12).

Nevertheless, notwithstanding the opposition, the truth is accessible. But we have also seen that perception of truth is commensurate with the desire for truth and for goodness. The fear of truth lies in what it requires of the believer. As the Prophet Joseph implied, man may fear confronting God: "If there be no resurrection from the dead, then Christ has not risen; and if Christ has not risen He was not the Son of God; and if He was not the Son of God, there is not, nor can be, a Son of God. . . . If He has risen from the dead, He will by His power, bring all men to stand before Him."[46]

Assenting to Jesus' resurrection requires acknowledgment of a powerful world of miracles. But that truth can be challenging, and when men find the truth too inconvenient or threatening, they indulge in deliberate blindness and denial, and they will censor truth's access to their conscious mind. They must practice self-deception to protect their position. But they pay a price, as resistance to truth always results in a general darkening of the mind (see D&C 84:54), a shutting down of one's own truth-discerning system, a slipping backward in evolving toward divinity.

Paul indicted those who cultivate darkness in the midst of continually manifested truth:

---

[46] Smith, *Teachings*, 62.

The wrath of God is . . . revealed from heaven against . . . men who suppress the truth . . . since what may be known about God is plain to them.

For since the creation of the world God's invisible qualities—his eternal power and divine nature—have been clearly seen . . . so that men are without excuse.

For although they knew of God, they neither glorified him as God nor gave thanks to him, but their thinking became futile and their foolish hearts were darkened. Although they claimed to be wise, they became fools (NIV Romans 1:18–21).

The quest for the historical Jesus need not depend on scant historical information in ancient records, or even on the testimonies in restored scripture, because "what may be known about God is plain."

In the fearless quest for the truth, obedience leads through the witness of the Holy Ghost to the full materialization of all things once held only in the eye of faith. Christ declares His accessibility: "*Every soul who forsaketh his sins and cometh unto me, and calleth on my name, and obeyeth my voice, and keepeth my commandments, shall see my face and know that I am*" (D&C 93:1; emphasis added).

But sooner or later the appointed hour will arrive when, with love or with fear, every eye shall see, every ear shall hear, every knee shall bow, and every tongue confess that Jesus is the Christ. There will come the moment when each person will receive his own resurrection and when every soul will look into the eyes of the Lord Jesus Christ and know with a perfect knowledge that He took His body again and lives as the resurrected Son of God and Savior of the world.